THE UNIVERSE AND THE CREATOR-GOD

by

John M. Mason

A Hearthstone Book

Carlton Press, Inc.   New York, N.Y.

Copyright © 1990 by John M. Mason
ALL RIGHTS RESERVED
Manufactured in the United States of America
ISBN 0-8062-3710-4

THE UNIVERSE AND THE CREATOR-GOD

CONTENTS

FOREWORD	BEFORE YOU READ	
INTRODUCTION	SCIENCE AND RELIGION	
CHAPTER I	LOOK UP AT THE STARS	17
	An orientation to where you are: on a Planet: in a Solar System: in a Galaxy: in a Universe.	
CHAPTER II	THE TURF OF THE SCIENTIFIC EVOLUTIONIST	26
	We must know the high points on which they stand in order to hold them to the pathway they have laid out for us.	
CHAPTER III	THE TURF OF THE WRITER	37
CHAPTER IV	THE BEGINNING OF THE UNIVERSE	50
	No one has ever seen an electron or a proton—but they are real—the stuff the Universe is made of. We know it had a beginning for it is moving on to its end.	
CHAPTER V	THE FORMATION OF THE GALACTIC SYSTEMS	70
	One hundred billion (or more) stars in a controlled system—orbiting the core of what we call a Galaxy. And in the Universe there are billions of Galaxies.	
CHAPTER VI	THE BIRTH OF THE SOLAR SYSTEM	90
	A small bit of the Milky Way Galaxy separated out from the arm of the Galaxy—to form a world with a diameter of ten billion miles.	

CHAPTER VII	THE PLANET EARTH, PART I: THE MYSTERY OF THE ELEMENTS	108
	The Big Bang Creation Cloud was made of gas—75% hydrogen and 25% helium plus a tiny smattering of heavier elements. The MYSTERY? Where did all elements 103 as found on Earth come from, and how?	
CHAPTER VIII	THE PLANET EARTH, PART II: THE MYSTERIES OF THE PROTECTIVE SYSTEMS	125
	The cosmic rays coming from outer space would burn up the Planet; the rays of the Sun striking directly would also burn it lifeless; the Solar Wind would smash it. But for the protective systems mysteriously built into the Solar System, the Planet Earth could not support life. How were they designed and put in place?	
CHAPTER IX	THE PLANET EARTH, PART III: THE FORMATION OF THE PLANET	136
	The force of gravity pulled the matter of the Solar System Cloud together. 99.9% of the mass of the Cloud formed the core, which became the Sun. The center of the core was solid iron, the next layer was composed of a mixture of liquid iron and nickel. The mantles, a plasma of molten matter came next and on top of the mantles floats the thin crust, the surface of the Earth. The nine planets, the Earth is the third one out from the Sun, and all the other bodies of the Solar System were formed from the 0.1+% of the mass of the Solar System Cloud. There are many questions.	

CHAPTER X	THE BEGINNING OF LIFE	150
	Life appeared when its time arrived—an intervention on the part of the Creator-God who supplied to the "life molecule" the quality of "aliveness" and provided the cell, the structure in which the life molecule functions. Life is a "given", the living spirit which controls all parts of the living cell, plant or animal.	
CHAPTER XI	LIFE AS WE KNOW IT	172
	There is no life without cells. In the adult human there are 60,000,000,000,000 cells! In the human 50,000,000 cells die every second but every second 50,000,000 new cells are born! Cells of plants and animals are almost alike—the difference being in the powerhouses of each. The plant absorbs energy from the Sun to provide energy for its own life and as a by product provides glucose which the animal cell translate into energy for its life system. In turn, it discharges carbon dioxide into the atmosphere for the use of the plant. Each life form is dependent upon the other.	
CHAPTER XII	THE GREAT INTERVENTION	199
	Charles Darwin in *The Origin of Species*, writes of the Creator breathing life into the first life forms. He credits "Natural Selection" and "Direct Action" (The Creator) for the development of all life forms since the beginning. At the critical mo-	

ments, the Creator-God intervened in the development of everything, including the human race! In the birth of Jesus, the Creator-God brought the human race into the eternal family of God. In his death he overcame Death; in his resurrection he announced his victory!

| POSTSCRIPT | THOUGHTS ON QUANTUM MECHANICS |

FOREWORD

The preparation for the writing of this book started about sixty-five years ago when I first began to study physics, chemistry and mathematics in high school. My teachers encouraged me in the inquisitive nature I showed and my parents indulged my efforts to create a scientific laboratory in our large attic. It is a wonder that I did not blow up the house or burn it down, or that I was not electrocuted as I conducted unsupervised experiments. My interest in the sciences continued as I studied at St. Olaf College in Northfield, Minnesota. These studies, however, led me into the field of theology.

Now that I have been retired as a clergyman for a number of years, I find that the same interests are with me. The past few years have been spent in reading the books of eminent scholars in the field of astronomy and astrophysics. Here I found that the scientific writers often drew theological conclusions as to the meaning of their findings but were very defensive of a dogmatic position relative to the non-existence of a Creator-God. The more I studied their positions the more I felt that a second opinion ought to be stated. This conviction led to the writing of this book, which I hope may be of help to many people who have had questions without answers.

In reading this book, you will find that when a scientific writer is quoted for the first time, a brief statement is made to identify his/her credentials. As it is anticipated that many readers may not be familiar with scientific terms and concepts, a pattern of repetition will be observed to help readers feel comfortable with the terms and understand the concepts more easily. The books and articles that have been used as resource material are listed in the bibliography.

INTRODUCTION

In writing this book I have not thought in terms of a continuing conflict between the two branches of knowledge, science and religion, for there should be none—each should support the other. This does not mean that there is no conflict, for that has been the unhappy story of mankind's quest for an understanding of himself and the Universe in which he lives.

Both sides in this struggle have been at fault, each claiming superiority over the other, each being dogmatic in their various pronouncements, neither being open to objective studies in which they would work together seeking answers to the same questions. In the last few decades there appears to have been a sort of truce, mutually agreed upon without open acknowledgment. There are good reasons for the truce, the first being that the conflict has been a total failure. No one has changed sides; the issues are still there, locked into position by the fact that the scientific community has tied itself to the rule of "the scientific method," allowing no argument to stand that cannot be proven in the laboratory. On the other hand, the religious community has taken an equally dogmatic stand to the effect that the Bible must be the proving ground for any proposal put forth by the scientists relative to the origin and development of the Universe as we know it.

A second and very significant reason for the polite talk between the two groups is that gaping holes have appeared in the armor of both parties. No one is anxious to risk the scorn of his peers by coming out with a flat statement to the effect that there are problems with both the "scientific method" and the "Bible Proof-Text Method." But the problems are there and they must be faced. This is the purpose of this book.

In these last few decades, when there has been a gigantic breakthrough on the part of the scientific community in its understanding of the Universe, both from the macrocosmic view and the microscosmic, there has also been a growing awareness that there cannot be two opposing Truths—one scientific and the other religious. Truth cannot be divided—what is true scientifically will not be found false religiously. Thus,the two groups have muted their criticisms of each other. As the scientists have traced the development of the Universe as we know it backward in time to a point only one-hundred-thousandth of a second after it all started with the Big Bang, they have become very uneasy. They fully realize, although they have not made this known to the public, that they have had to put the scientific method on hold because it just is not possible to apply it to all the hypotheses they have had to work through in order to arrive at that critical point when the Universe exploded into being.

Eugene Mallove, an astronautical engineer, wrote in *Commentary (Minneapolis Star and Tribune*, Nov. 4, 1985) about the puzzle scientists have over the inexplicable coincidences they find in the cosmos. "Some scientists are awestruck by the numerous improbable physical coincidences in the universe, without which life could not exist. Their work is forcing them to ask the ultimate philosophical question, long the province of religion: Did the Universe come into being by chance or by design?"

Hoimar von Ditfurth, professor of psychiatry and neurology and an award-winning author of popular scientific books in Germany, has come a long way toward acknowledging the necessity to handle the scientific method with care. In his typical scientific language, he says in *The Origins of Life*, "The refutation of the principle of verification (the demand that we take seriously only the things whose truth can be positively demonstrated) has liberated us from the positivistic gag rule." In other words, we can live with the scientific method if we realize that it is not absolutely binding in all circumstances!

To accept the concept that the Universe came into being by way of a great explosion of energy raises questions the scientific method cannot handle, such as: What was the source of the energy in the pre-Big Bang fireball? Where did this energy come from? How did "laws of nature" come into being? The questions sound just like the old ones about the existence of God! Where did he come from? What moved him to create the Universe? How does he govern the Universe?

It is obvious that the scientist cannot prove that God does not exist. Hoimar von Ditfurth states this boldly on page 183 of *The Origins of Life*: "We have modern science to thank for the knowledge that the possibility of God's existence and of a transcendent reality can in no way be refuted." As scientists have moved toward the macrocosmic view of the Universe, a development of concepts occurred through the acceptance of hypotheses which cannot be tested by way of the scientific method. The same is true, and probably much more so, as they discover the mysteries of the microcosmic realities of the Universe. Science has become unsure of itself as it struggles with the implication of the probability theories in quantum mechanics. The scientific method was a useful tool along the way—but for now it is partially "on hold."

And so also with the theologian: the breakthrough in knowledge about the origin and development of the Universe has not gone unnoticed. Some Bible believers closed their minds to the findings of the scientists and adopted a position in which the "Bible Proof-Text Method" was final. They called themselves religious fundamentalists—whatever that word might mean. Many more, however, quietly went about the task of examining the findings of the scientists, accepting some conclusions and holding back on others. The tragedy is that they were not ready to examine openly with the scientists those areas in which there were problems. The great difficulty that has confused the issues is the violence with which the fundamentalists have held to their position in opposition to all the evidence which speaks to the age of our Planet Earth, the age and development of plant and animal life on Earth and the place of the Planet in the Solar System and the Solar System in the Milky Way Galaxy. The stumbling block that appeared to command the greatest attention of the fundamentalists was the doctrine of evolution. The name of Charles Darwin was held up as the perverter of all religious truth about the origin of mankind. It is doubtful that much thought was given by the Fundamentalists to the evolution of plant and animal life—to say nothing of the evolution of inorganic matter, all of which is part of the total story of evolution being the creative process used by the Creator-God in bringing his Universe as we know it into being. Most attention was riveted to the importance of denying the possibility that human life evolved from the ape—just a small part of the real story.

The religious community must take responsibility for not having examined the statements of Charles Darwin more carefully. The theologians should also have noted that in his first issue of *The Origin of Species*, Darwin made a strong complaint that his position on "natural selection" and the effect of "direct action" were being misrepresented by the media. This misrepresentation is still going on. Darwin did not teach that life began spontaneously in a warm pond in the shallows of the primordial sea. It may well be that such a warm pond cradled the first life form into which, Darwin said in the closing paragraphs of *The Origin of Species*, "The Creator breathed life!" Darwin did not teach that evolution occurred randomly by "bizarre genetic mistakes" as the scientific evolutionists say, but rather that the development (evolution) of life and of the species came about by "natural selection" aided by "direct action" on the part of a source "that we in our ignorance do not know."

In other words, Darwin taught that life came into being by the direct action of the Creator-God who breathed life into the first life forms, and that subsequently life developed as these first forms multiplied and changed by the process of natural selection under the laws of nature that were in place and by direct intervention on the part of the Creator-God who was continuously involved. Darwin also made very plain in the final paragraph of *The Descent of Man* that the whole evolution process was not carried out by "blind chance," but was a planned and ordered evolution of life according to the sustained, expressed will of the Creator-God.

While it has been made clear, at least by von Ditfurth, that in some areas the scientific method is on hold, we must not fail to recognize that it is basically a valued tool for both the scientific and the religious. The human race has a built-in sense of the need for God; this is a genetic characteristic that is not found in plants or animals as far as we know, and separates mankind from other life forms. This is plainly spoken of in the Bible: "I will put my laws into their mind and on their heart also will I write them. And I will be to them a God and they shall be to me a people" (Hebrews 8: 10). And again we quote von Ditfurth: "At all times and on all continents, in all cultures and phases of history, humans have been religious, have either believed in the existence of a reality beyond this world, or at least considered it a serious possibility." This "God-Need" built into the genetic code of humans, separating them distinctly from all other life forms, constitutes empirical evidence for the reality of the existence of the Creator-God. Darwin would say that it is an example of "direct action" from an outside force that long ago began the process of making man in His own Image. He is not finished yet—the process of evolution is still in operation!

Thus, in the futile struggle between the scientific and religious communities, we see that both sides have problems with their basic historical positions. The time has come for mankind to face these issues openly and without the dogmatic stance that has so far characterized the conflict.

Now that we have come to a point in time when the scientific and the religious are both seeking a way out of dilemmas of their own making, we should be pleased that a science writer of the stature of von Ditfurth has produced a kind of "transition" document in his book, *The Origins of Life*. We—I speak for at least some of those in the religious camp—are grateful for this book. We hasten to warn, however, that we must not be misled. Ditfurth has moved part way but is far from the point of having accepted a concept of the existence of God that would allow for a personal God, concerned about the life of people who have been created in His own likeness.

In other words, the caveat is simply this: The original struggle between science and religion was one in which the ground rules had been laid down by the scientists in the adoption of the scientific method. The religious reacted to this by setting up the Bible Proof-Text Method and we have been at an impasse ever since. This must not be allowed to happen again! Ditfurth's fine book, much of which I can accept without question, is exactly the vehicle for a new, science-dictated set of ground rules. As he says in his Introduction: "And so I offer this book to the churches as an attempt to explain, from a scientific perspective, the ways in which science and religion might after all function harmoniously together and to sketch out some possible avenues of rapprochment between the two."

And so I would be so bold as to offer *The Universe and the Creator-God* to the churches and to the scientists in the hope that a beginning may be made in laying the groundwork for building a right relationship between Science and Religion.

THE UNIVERSE AND THE CREATOR-GOD

CHAPTER I
LOOK UP AT THE STARS

On a cloudless night when the moon is down, stand on a high point and look up at the stars in the sky. What you see in those sparkling lights is a tiny part of the great universe in which we live. Let your mind travel to those distant suns and realize that the light beams coming to you began their journey to the Planet Earth years ago. If you should happen to see Alpha Centauri, which is the star closest to our earth, know that its light as you see it began its long journey to your spot about four and a half years ago. At 186,200 miles per second, it has traveled 26,442,113,040,000 miles through space to reach you on your high point! We could write that figure as 26 trillion, 442 billion, 113 million and 40 thousand miles. And this is our nearest star!

It is difficult for our minds to conceptualize what a great distance that is. Our Sun, by comparison, is only 92,900,000 miles from Earth. It takes a light wave only 8 minutes to travel from the surface of the sun to the surface of our planet. The Solar System, which is controlled by the sun, consists of nine planets, many moons, asteroids, comets and clouds of dust and debris. The Solar System can be likened to a sphere within which all of these bodies are in orbit around the sun. It is held together by the strong gravitational pull of the sun in such an orderly manner that the planets keep their proper places in their orbits so accurately that we can predict when there will be an eclipse of the moon (when the earth lines up between the sun and the moon) or an eclipse of the sun (when the moon lines up between the earth and the sun). We keep our time by the movements of these heavenly bodies in their orbits. We know when the great comets will come into view and when we will be favored with the fireworks displays which we call "falling stars." The constellations of stars in the heavens can be counted on to be in their places at the proper

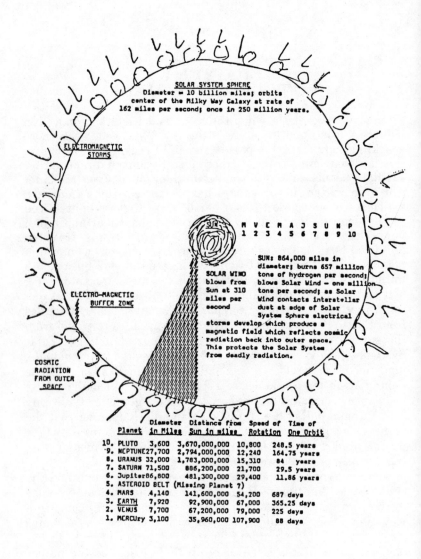

times and our days, months and years, determined by the position of our planet relative to the sun and moon, can be double-checked by those bodies outside of the Solar System. There are more precise time-keeping systems, like the atomic clock, but for everyday use we get along very well with our solar system clocks.

The sphere of the solar system—diagrammed in Figure 1, not to scale—has a diameter of about 10,000,000,000 miles. The tiny planet Pluto, which lies near the outer limits of the solar system, is almost 4,000,000,000 miles from the sun! As our solar system has measurable dimensions and follows a specific orbit around the center of the Milky Way Galaxy, we must think off it as an organized body, moving and functioning in accord with the laws of physics governing bodies in space. It appears, as we will see later, to have a border zone around its outer limits consisting of a magnetic field which acts something like a shield, protecting the solar system against the highly active cosmic radiation, strong enough to destroy all life on the Planet Earth. We will also see that all special bodies outside our solar system and even beyond our Galaxy, follow the same rules or laws. Perhaps we should hedge a bit at this point to mention that some scientists are puzzled by what they seem to observe in the far out galactic spaces, out toward what appears to be the limits of the Universe, or beyond 15 billion light years. (A light year measures the distance—about 6 trillion miles—traveled in one year by a light wave, which moves at 186,200 miles per second.) It has been suggested that possibly a different set of laws of physics prevails in those distant regions. It is doubtful, however, that if the laws show themselves to be constant in all the space between, they would suddenly function differently. Rather, it may well be that our instruments cannot adequately report what is actually happening in those far-out regions. As our instrumentation is improved, we likely will get the true picture. We should understand, however, that this tells us that the distance from our high point here to the "edge" of the Universe must be about 90 billion trillion miles!

We have mentioned the Milky Way Galaxy. When you stand on your high point, you will see that a wide band of stars in the sky seems to be cloudy, hazy or milk-like. The word *galaxy* comes from the Greek word for milk—galaktos—and the ancient astronomers named that cloudy area in the heavens, the Milky Way.

It is also an organized body in space, made up of more than one hundred billion stars—our sun is one of them—plus a great amount of hydrogen gas, other gases, clouds of dust laced with newly formed elements, and interstellar debris, all in orbit around the gravitational center off the galaxy. Our galaxy has been in existence since the time of the break-up of the original Creation-Cloud off radiation which followed the Big Bang of Creation more than fifteen billion years ago!

Our Galaxy is called a *spiral galaxy* for as it turns on its axis there seem to be huge arms trailing around it, as illustrated in Figure 2. The huge number of stars, planets, gas clouds and other materials in the galaxy provide us with the diffused light which gives the impression of a milky band in the sky. Our Sun is located between two spiral arms, about as shown in the diagram, not in the center of the galaxy, for no life could exist in a place so hot, but only in the flattened area nearer the edge. We should remember as we look "up" at the Milky Way that we are looking at it from the "inside" and thus have a distorted view. We don't see it quite as it is. To us, it looks like something not only far away but "apart" from us when actually we are inside of it! Hoimar von Ditfurth, a scientific writer and a professor of psychiatry and neurology at the University of Wurzburg and the University of Heidelberg, in his book, *Children of the Universe*, states, "With the naked eye we can see less than 1/10,000,000th of the 100 billion stars in our galaxy" (p. 39).

Our galaxy is not alone in space. There are billions of them! Some are larger than the Milky Way while many are smaller. Our galaxy is looked upon as being of average size even as our sun is an average-sized star. The Milky Way is, nevertheless, 100,000 light years in diameter, or 600,000 trillion miles across. The average distance between galaxies is believed to be about 20 million trillion miles. There is really little danger that one galaxy will collide with another—and should such a thing occur, as we will see later, perhaps nothing much would happen! The movement of these giant bodies of organized units in outer space appear to be controlled in some manner; they move in a somewhat orderly pattern, which very likely makes a collision highly improbable. There are other kinds of galaxies in outer space, such as the elliptical and the irregular. We will describe them at another point. They are as large as the spiral type of which we are a part.

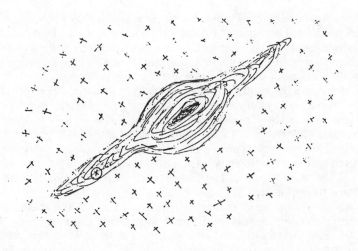

THE MILKY WAY GALAXY

A typical "spiral" galaxy—discus shaped—about 600 thousand trillion miles in diameter (100,000 light years). It has upwards of 100,000,000,000 stars—of which our Sun is one! Its approximate location is indicated by the X. Our star, with its Planetary System orbits the gravitational center of the Galaxy once in about 250,000,000 years—traveling at a speed of 162 miles per second. The Galaxy is moving "outward" as the Universe continues to expand. The center of the Galaxy is 90 trillion miles in width (15 light years) and consists of a huge nuclear furnace which is the gravitational center of the Galaxy. The bulge area is about 18,000 trillion miles in width (3,000 light years). Besides the stars, there are huge clouds of hydrogen in the Galaxy—which have an important role in new star formation. There are also clouds of dust, containing carbon, nitrogen and chemical compounds—likely the products of exploding Supernova Stars. Around and in the space of the Galaxy, roughly in the shape of a sphere, are groups of stars, called Globular Clusters. They are shown here as X's. It is believed that they are the first stars formed after the Galaxy Cloud separated out from the Creation-Cloud—before the Galaxy had taken on its present discus shape. Each cluster is about 1,200 trillion miles in width, containing possibly millions of stars. More than one hundred Globular Clusters have been identified in the Milky Way Galaxy. Other Galaxies also have them—our nearest Galaxy, the Andromeda, has more than two hundred. These stars present a mystery which we will touch on in Chapter VII.

The spiral galaxies appear to be very similar to each other in composition, consisting of from fifty to several hundred billions of stars. Some of the stars may well have planets, solar systems, moons, asteroids, cosmic dust, gas clouds and debris, in much the same fashion as we have described for our solar system. In fact, astronomers have come to agree on a *Cosmological Principle* that the Universe is everywhere the same. Paola Maffei, professor of astronomy at the University of Catania and director of its astrophysical observatory, in his book *Beyond the Moon*, puts it like this: "This concept expresses the so-called cosmological principle, which stated in its most condensed, precise form, says: On a large scale the structure and the properties of the universe are everywhere the same" (p. 317).

As you stand on your high point, looking up at the stars, marvelling at their beauty and wondering about their mystery, try not to lose your balance! You also are in orbit in ways that would be difficult to explain. You are rotating with the Planet Earth as it turns on its axis. You make one turn around in 24 hours, which defines your day. At the equator, where the circumference of the earth is about 25,000 miles, the speed of rotation is more than a thousand miles an hour. For those who live north or south of the equator, the speed slows down according to the distance from the equator. At the two poles the speed approaches zero. Our planet is also tilted on its axis, which gives us our four seasons. When the light of the sun strikes us more directly, we have our summer. In the winter, the angle of the sun is changed so that the light strikes us less directly—meaning that the same amount of sunlight is spread across a greater area and thus the temperature is lowered. In the spring, we are moving toward the summer setting and on June 22 we are at what is called the summer solstice, a word that could be translated "when the sun stands still." Actually, it marks the time when the sun is farthest out from the equator and is about to begin to move back toward the equator. We call that day the first day of summer. In the fall the opposite happens, and on December 22 we have the winter solstice, the beginning of winter, when the sun again changes direction relative to our place in the Northern Hemisphere. In the Southern Hemisphere, it is just reversed; their summer solstice is marked on December 22 and their winter begins on June 22. We have two other points to mention, the equinoxes. This word could be translated as meaning "equal nights." The time when the sun crosses the plane of the earth's equator, making night and day of equal length, is called the equinox. In the spring we have the vernal equinox falling on March 22, the first day of spring. The autumnal equinox falls on September 22, marking the first day of fall. The tilt of the earth, at 23½°, is of great importance to us. If it were not just as it is, our summers could be too hot and our winters too cold. Very likely life would not be possible on the earth, or would be much different than it is.

As you speed on your journey in what we call a west-to-east direction, you are also moving with the earth in an orbit around the sun. This turn will take 365.25 days, which defines our year and you will be traveling at the rate of about 67,000 miles an hour! The trip will cover about 67,000 × 24 × 365.25 or about 587,322,000 miles. And this is not all. The entire solar system is turning on its axis and taking us along with it. Then the solar system with all of its planets, moons, satelites, gases and dust, is in a gigantic orbit around the gravitational center of our Milky Way Galaxy. Scientists tell us that we are traveling in this direction through space at the rate of 162 miles per second! It will take 250,000,000 years for you to make this trip and be back at your high point of observation! You would not, of course, be in the exact spot in space but you won't know the difference. In making this orbit, the Planet Earth, with all of its companions, will have traveled about 162 × 60 × 60 × 24 × 365.25 × 250,000,000 = 1,253,082,800,000,000,000 or one quintillion, 253 quadrillion, 82 trillion and 800 billion miles!

In still another direction (we use the term lightly for by now we don't know what it means), our galaxy is turning on its axis and it also is either in orbit around the center of the Universe or it is moving "out" from wherever it started—scientists are not in agreement about this. We do know that there are billions of galaxies like our own and that they are in motion similar to that of the Milky Way. The galaxies are moving at different speeds, depending on how far they are from wherever they started! Some have been measured at the incredible speed of a thousand million miles per hour! They also appear to be moving away from our galaxy and the farther away they get, the faster they move. The suggestion that our galaxy, or even our solar system, must then be the center of the Universe, is generally not accepted by the scientific community, which claims that all the galaxies are moving away from each other and that if we were located on one of the other galaxies, we would think that all the others were in motion away from whatever galaxy we were on. At this point it should be noted that, if this is true, the galaxies appear to be breaking the pattern of the other moving bodies in space. For instance, the earth revolves on its axis and orbits the sun; the sun revolves on its axis and orbits the center of the galaxy. The galaxy revolves on its axis and it really ought to orbit the gravitational center of the Universe. The matter remains a mystery, for the explanations of this phenomenon fall short of being convincing. For everything in the Universe to be moving away from everything and at differing speeds, appears to be a logical absurdity unless there is a defined center, which scientists do not want to accept. The Big Bang explosion must have occurred "somewhere" and that should provide the "point of origin" for the expanding Universe. There would appear to be no reason why orbiting galaxies could not also be in a movement which would be enlarging the size of the orbit constantly, that is, by moving "out" and at an increasing speed as they move farther from the gravitational center of the Universe. We will deal with this matter again.

The galaxy nearest to our Milky Way is called the Andromeda Galaxy. It is two million light years away from us or about 12,000,000,000,000,000,000 (12 quintillion) miles. Thousands of galaxies are within one hundred million light years and billions are within the range of the 200-inch telescope on Mt. Palomar. For anyone to attempt to comprehend the total volume of space in which the Energy-matter of the Universe moves, would be to invite frustration. As new equipment and technology come into being, the outer and inner limits of the Universe become very complex. When we look at how huge the Universe is we are taking the macrocosmic view and seek to determine how small the constituent parts of the Universe are. This we will look at when we seek to understand the make-up of the atom.

One thing appears to be certain, without question. We are! We exist on a Planet, in a Solar System, in a Galaxy, in a Universe. We are learning more about each of these categories and will learn much more as time passes and the already sophisticated instruments continue to be improved. Scientists the world over have come to an agreement on a point, uncomfortable to many, but one which is based on evidence in hand and thus must be accepted. They agree that the Universe has not existed forever, that it came into being in an instant when a huge mass of Energy-matter composed of electrons, protons, neutrons, photons and neutrinos—or their constituent parts (!)—exploded into space. All of this enormous mass of energy was compressed into a fireball with a temperature of about 100,000,000,000° C. Maffei, in *Beyond the Moon* (p. 322), puts it in these words: "At that time all the matter and energy in the Universe must have been concentrated, at the enormous temperature of 100 billion degrees, in a sphere that astrophysicists have named the fireball. This sphere had a radius scarcely one ten-billionth that of the present Universe and was composed of protons and neutrons, radiation of extremely high density, and a gas of photons, electrons and neutrinos. Although it was in thermodynamic equilibrium, this fiery mixture could not remain static for an indefinite length of time (as the astrophysicists have shown); thus, at a certain moment, it must have begun to expand. It is at this point, with a change taking place, that we can begin to speak of time."

The scientific evolutionist tells us that this explosion marked the beginning of time, that there was no "before" and that when all the energy of the Universe has burned out, there will be no "after". Robert Jastrow, founder and director of NASA's Goddard Institute for Space Studies, in his book, *Until the Sun Dies*, states: "The latest answer is that all the matter and energy, in these many forms, will not suffice to bring the expansion of the galaxies to a halt. According to the available facts, the material of the Universe must disperse forever, until all is space and emptiness. It appears that there was only one beginning and there will be only one end." He had earlier in the same book written, "The lingering deadline predicted by astronomers for the end of the world differs from the explosive conditions they have calculated for its birth, but the impact is the same: modern science denies an eternal existence to the Universe, either in the past or in the future."

If you are still standing on your high point, looking up at the night sky with its beautiful array of sparking lights, do not give up hope. While we do not deny the factual evidence of the scientists concerning the life story of the Universe, we believe that some unjustified conclusions have been drawn. It is our purpose in this book to examine as well as we can the evidence the scientists have given us. In doing this, we will hold the scientific community strictly to its own turf, the ground on which it has taken its stand.

CHAPTER II
THE TURF OF THE SCIENTIFIC EVOLUTIONIST

The scientific evolutionist has taken his stand, he has staked out his turf. The lines have been drawn clearly. He has chosen the ground on which his hypothesis is built. He believes that at no time has any outside force or intelligence had any part in guiding or directing the course of events which have led to the origin and development of the Universe. He believes that it has happened all by chance. It must be said forthrightly that he has done a splendid job in tracing the events from the time just a fraction of a second after the Big Bang of creation down to the present. Those who would doubt his astounding findings would do well to read some of the many volumes which have been written by members of the scientific community in the past two decades. The list in the bibliography for this book are suggested.

A basic error which the scientific evolutionist has made, however, is to confuse description with explanation. No matter how carefully and accurately he may describe what has happened in the vast expanse of time since the Universe came into being, the description does not explain the origin of the Universe nor how its development was achieved.

Thus, while we might accept for real the discoveries which have been made about the Universe in which we live, there is room for a second opinion on the conclusions that have been drawn. It is for this reason that this study is made. It is hoped that by the acceptance of the astrophysicists' findings as well as those of other scientists, and by reckoning with a Creator-God, the work of the scientists may be appreciated fully and the meaning of life in our Universe may be enriched.

In fact, if this cannot be done, life as we know it can only end as a tragedy. This is expressed most poignantly by one of the great contributors to the cosmological knowledge of our day, *Dr. Steven Weinberg*, Higgins Professor of Physics at Harvard University and Senior Scientist at the Smithsonian Astrophysical Observatory, in his excellent book, *The First Three Minutes*. He expressed his despair in the last paragraph of his book. It is with deep sympathy for him and indeed for all of his fellow scientists who hold his view, that we say that his story ends with a sob. He wrote: "But if there is no solace in the fruits of our research, there is at least some consolation in the research itself. Men and women are not content to comfort themselves with tales of gods and giants, or to confine their thoughts to the daily affairs of life; they also build telescopes and satellites and accelerators, and sit at their desks for endless hours working out the meaning of the data they gather. The effort to understand the universe is one of the very few things that lifts human life a little above the level of farce, and gives it some of the grace of tragedy."

But human life does not have to be lived in such misery. Many times, as I have read and reread the writings of these men and women who have dedicated their lives to gaining an understanding of the Universe, I have thrilled to the realization of the wonder of it all. The Laws of Nature, as mankind has been able to deduce them from the evidence in the Universe, with an occasional assist from the One who is the Law Giver, clearly shows, not only the order of the Universe, but the beauty, the magnificence of it, and the loving care which the Creator-God has bestowed upon it. Otto R. Frisch, Jacksonian Professor of Natural Philosophy, Cambridge University and Director of Research in high-energy physics at the Cavendish Laboratory, wrote in *The Nature of Matter* about the "assist" that comes to the dedicated worker: "But important advances invariably come by virtue of an inspired guess, a sudden flash of insight; and in this respect the work of the scientist does not differ so much from that of the writer, painter or composer" (p. 10).

Nevertheless, the scientific evolutionist has taken his stand. Very clearly he has chosen to risk everything on a hypothesis which by his own testimony he cannot prove. Dr. John Gribbin,

Doctor of Astrophysics, Cambridge University, in his discussion of the Big Bang of Creation in *Genesis: The Origins of Man and the Universe*, states on page 32, "How the very dense, very hot Universe came into being we cannot, in all honesty, say. We can, however say how it got from being very hot and very dense to being almost empty and very cold, the state it is in today." Dr. Robert Jastrow states in his book, *Until the Sun Dies* (p. 28), "A sound explanation may exist for the explosive birth of our Universe; but if it does, science cannot find out what the explanation is. The scientist's pursuit of the past ends in the moment of creation; the origin of the world is a fact that he can never hope to explain."

The scientist has chosen to build his case for the existence of the Universe on a concept which suggests that everything that we can observe, weigh and measure, everything that we are, all that we know, all Energy-Matter, has come into being and has developed to its present state by happenstance. He believes that out of the primordial cloud that formed in space immediately after the Big Bang, the random collisions of protons and electrons formed the matter which in the period of billions of years developed the Galaxies of the Universe, the stars in the galaxies, the Solar Systems in the realms of the stars, the Planets in the solar systems and finally on one Planet, at least, life as we know it—even the finely tuned instrument we call the human body with more than 60,000,000,000,000 (60 trillion) cells.

Jastrow in *Until the Sun Dies* states: "Undismayed by his failure to explain the moment of creation, the scientist has undertaken another bold endeavor; he has undertaken to explain the existence of man by scientific methods as straightforward as the calculation of the orbit of the planet Mars. And he has nearly succeeded; here, in place of a miracle, he can offer a chronicle of events by which simple elements in the parent cloud of the Universe gradually evolve into conscious life " p.51). Note that here also he has only "nearly succeeded"; the mystery of the origin of life still eludes him.

The puzzling thing about the stand of the scientist is the obvious fact that he has lost sight of the most sacred principle that has governed science for the last few centuries—the scientific method. Hoimar von Ditfurth, in *Children of the Universe*, states: "The world view we now designate as medieval did not survive the advent of the scientific method" (p. 9). This method is one in which the scientist carefully examines and tests the empirical evidence for every hypothesis under controlled laboratory conditions before accepting it as a theory, and especially as a scientific fact. It has been a very useful tool and no doubt has done much to advance our knowledge of the Universe and its constituent parts. No one would dispute that the scientific method has contributed so much to science that it has justified its place in the history of mankind. Now, however, in the light of the conclusions being drawn by the scientific community relative to the formation of the Universe and its subsequent development, it would be fair to state that the scientific method has not survived the speculations of the modern scientists. In the development of the current views on the Universe, the scientist has rejected the impirical evidence which indicates that the Universe is not just a happenstance accident.

The scientist's insistence on his proposal that everything has developed by chance is in itself a rejection of the scientific method. In seeking to establish his case, he has in an amazing manner adopted a vocabulary which cannot be accepted as scientific. He uses freely words like "somehow," "perhaps," "by accident," "by chance," "probably," "randomly," "willy-nilly" and even "magic"! These expressions do not conform to the usual carefully calculated words used to describe a new scientific theory. The evidence of the abandonment of the scientific method will build up as we move on in reviewing the story of the scientific evolutionist, as told by the scientists themselves. At this point, in defining the turf of the scientist, we are simply stating that we will hold him to his position that it all has happened by accident.

The obvious, which the scientist has overlooked—or more likely, refused to look at—is that he has not accounted for the atom. Jastrow quite accurately describes the reaction of the scientific community to the fact that the Big Bang dictates that the Universe had a beginning. He wrote in *God and the Astronomers*: "When evidence uncovered by science itself leads to a conflict with the articles of faith in our profession . . . it turns out that the scientists behave the way the rest of us do when our beliefs are in conflict with the evidence. We become irritated, we pretend that the conflict does not exist, or we paper it over with meaningless phrases" (p. 16). Thus, the men and women of the scientific community ignore the fact that the key to the explanation of the existence of the Universe lies in the "coming into being" of the hydrogen atom. This, as we shall see at a later point, is a far more complicated matter than the simple orbiting of a negatively charged electron around a positively charged proton. Von Ditfurth in *Children of the Universe* states: "In the beginning was hydrogen. Besides hydrogen, nothing else existed but the laws of nature and space itself—unimaginable quantities of space" (p. 283). It is interesting to note that he acknowledges the existence of the "laws of nature" but does not recognize that there must have been a Law Giver.

Jastrow in *Until The Sun Dies*, states: "These pockets of gas that evolve into stars *are formed by accident by the random motions of the clouds that surge and eddy through the Universe*" (p. 52, emphasis mine). He is writing about the hydrogen gas clouds that emerged from the original fiery mass of radiation that burst into being at the time of the Big Bang when the Universe was born, according to the modern scientific view. He apparently does not want to deal with the obvious fact that the existence of whatever it was that exploded must be accounted for. Like von Ditfurth, he avoids any explanation of how the sub-atomic particles of energy that formed the hydrogen gas had come into being when the temperature had cooled to a point when a stable atom could exist.

It must be remembered, as we make this study, that all the Energy-Matter of the Universe as we know it, the billions of galaxies, each of which contains billions of stars many of which may have solar systems similar to ours, all of the white dwarfs, the black dwarfs, the red giants, the neutron stars, the quasers, the black holes and whatever else there is in the Universe, including all the clouds of gasses which "surge and eddy" in the vastness of space, must have been contained in that huge fireball when it exploded.

In *God and the Astronomers*, Jastrow states: "At this stage all of the Universe that we can see today was packed into the space of an atomic nucleus. The pressure and temperature were also extremely high and the Universe was a fiery sea of radiation, from which particles emerged only to fall back, disappearing and reappearing endlessly (p. 123). At this time, if we can call it "time," the density of the fiery sea of radiation was 10^{90} tons per cubic inch and the time was 10^{-43} seconds after the Big Bang. To write these figures out helps us to realize that we do not have words to describe them—1,000,000,000,000,000,000,000, 000,000,000,000,000,000,000,000,000,000,000,000,000, 000,000,000,000,000,000,000,000,000,000 tons per cubic inch! It has been calculated that in one second the temperature, which was 1,000,000,000,000 (one hundred billion) degrees at the time of the explosion, had dropped to ten billion degrees. Jastrow continues to state that the building blocks of matter—the electrons, protons and neutrons, and their anti-matter counterparts and the other energy particles such as photons and neutrinos—began to form as the temperature dropped. They "condensed out of the sea of hot radiation like droplets of molten steel condensing out of the metallic vapor in a furnace" (p. 126). When the temperature had dropped to ten million degrees, at about three minutes after the explosion, protons and neutrons began to form the nuclei of the helium atom. The scene was being set for the time when the temperature would allow the electrons to begin to orbit the protons to form the hydrogen atom, which marks the beginning of the existence of matter in the Universe. It is the process of matter-formation out of energy particles which the scientific evolutionist conjectures to have been a happenstance incident. But without an explanation of the formation of the fireball itself, it is presumptuous on the part of the scientific community to simply state that all of the Universe was formed by the accidental collisions of hydrogen atoms in the random movements of the huge clouds of hydrogen gas that were thrown out into space at the time of the Big Bang.

In his introduction in his book *Genesis—The Origin of Man and the Universe,* John Gribbin states: "In this book I have attempted to provide, for a reader with no background in science, an overview of the best modern, scientific answer to the question 'Where do we come from?'" He goes on to say, "In the beginning there was nothing at all. This is a very difficult concept, and one which causes a great deal of misunderstanding among many people who have heard of the idea of the 'Big Bang', the creaation of the Universe as we know it in some vast explosion of matter and energy....But before the Big Bang of creation, there wasn't even any empty space. Space and time, as well as matter and energy, were created in the 'explosion,' and there was no "outside" for the expoding Universe to explode into, since even when it was just born and beginning its great expansion, the universe contained everything, not just matter and energy, exploding into the void, but the void itself—space. And not just space, but its counterpart, time, the other facet of the space-time fabric. The flow of time as we know it also began with the Big Bang so that it may be literally meaningless to ask what happened 'before' the Big Bang—perhaps there was no 'Before'!"

Like his scientific counterparts, Dr. Gribbin refuses to recognize the obvious, that nothing can "explode" if something does not exist! He goes even farther to state that there was not even any space in which the explosion could occur. There was thus no "point" of explosion. It is difficult to accept this position from a logical point of view and doubly difficult when at other times the scientists speak of the "expanding" Universe, which certainly implies "expansion from some beginning point." The concept of time beginning at the moment of the explosion is acceptable in the sense that as we measure time, we can do so only in relationship to the beginning of the Universe. Certainly the possibility of a "Before" is not ruled out simply because time as we know it began when the Big Bang took place. In the quote just mentioned, Gribbin does not recognize the possibility of the "Eternal" nor the fact that within the concept of the Eternal, Space as such was a reality. Universe time is relative only to Universe happenings but apart from Universe-related happenings there is the Eternal Existence involving the Creator-God and all else that may be!

Jumping ahead billions of years (for the moment!), we quote Jastrow again from *Until The Sun Dies*: "Now and then collisions occur between neighboring molecules in the broth; in some collisions the small molecules stick together to form a larger one; then another small molecule collides and sticks, and still another. In this way, during the course of a billion years, every conceivable size and shape of molecule is created by *random* collision" (p.59). This is truly an astonishing statement coming from an outstanding scientific writer. Even if every conceivable size and shape of molecule would have been formed accidently in this manner, they would constitute only an unorganized mass of molecules, not a highly organized Universe such as we know. If all that is has come into being by the random collisions of atoms there can be no "laws of nature." The concept that accidents which happen randomly and yet consistently result in the development of polyatomic structures which can be described mathematically and tested in the laboratory, does not conform with the concept of "laws of nature," the laws of physics and the will of the Law Giver. Thus, the concept that the Universe came into being and developed to its present state by sheer accumulations of random happenings which have occurred in the vastness of space and in the centers of stars, without design or purpose cannot stand.

Gribbin has also taken his stand on the happenstance aspect of the development of the Universe and the life in it. In doing so, he has aligned himself with other respectable scientists in backing the hypothesis that everything from the time of the Big Bang to the present has happened by chance. He states in *Genesis* (p. 189), "Over many millions of years, the accumulation of such rare beneficial copying errors (mutations) gives rise to species as diverse as a mouse and a mushroom. *But the process happens willy-nilly.* and in the replicating game the single-celled varieties that have been unchanged for thousands of millions of years could, from one point of view, be regarded as more 'successful' than *the collection of bizarre mistakes* that has produced you and me."

Returning to Jastrow and *Until The Sun Dies*, we read: "Eventually, after *countless millions* of chance encounters, a molecule is formed that has *the magical ability to produce copies of itself* (p. 59). Whatever happened to the scientific method? When a scientist resorts to magic to make his point, anything can be proved. Again, in the same connection he wrote, "In a short time, they [the molecules that reproduced by magic!] dominate the population of molecules in the waters of the young earth." If this be so and if this be the rule of the scientific evolutionist, we should not now be painstakingly combing the crust of the earth to locate the "missing links" between animal and human life. There should be great numbers of skeletons available to demonstrate the subtle changes which we are told took place while moving the ape to the human state. Actually, there is a great gap in the fossil record of the scientific evolutionist with respect to the development of the hominoid. Again, we quote Gribbin, the scientist, in *Genesis*: "Frustratingly, though, there is a gap in the fossil record from the time of the Ramapithecus (10 Myr BP) [ten million years before present] up to about 5 million years ago" (p. 282). He goes on: "There are no fossil hominoid remains yet known for the period from 10 million years ago to 5 million years ago, and then nothing except one jaw fragment which experts are still arguing over, until 3 million years ago. The experts don't always like admitting how little we know even about Ramapithecus—just a few fossil fragments, identified as an ancestor of the human line from the jaw shape" (p. 289).

Huston Smith, professor of religion at Syracuse University, New York, writing in *The Christian Century*, July 1982, in an article called "Evolution and Evolutionism" states: "On a different front, with the displacement of Darwin's gradualism by the 'punctuational' model, it is now conceded that the 'missing links' between most species will not be found. It all happened too fast." He then quotes Steven Stanley in "Darwin Done Over": "Most change has taken place so rapidly and in such confined geographical areas that it (evolution) is simply not documented by our imperfect fossil record" (*The Sciences*, Oct. 1981). We will deal with this subject again at the time when change by mutation is discussed.

We must return now to what Jastrow writes concerning the beginning of the Universe. He writes in *Until the Sun Dies* of "this great saga of cosmic evolution . . . without a known cause" (p. 21). We must look carefully at what the scientists have said happened just after the explosion of creation in order that we might understand as clearly as possible how they believe the Universe was born and the great galactical system was formed. Jastrow writes: "Only as a result of the most recent discoveries can we say with a fair degree of confidence that the world has not existed forever; that it began abruptly, without apparent cause, in a blinding event that defies scientific explanation" (p. 19). As we take this look, we must keep in mind that the scientific evolutionist must be contained on his chosen turf, the ground on which he has taken his stand: that the Universe came into being suddenly with the explosion of what he has called a fireball in an event which he has named the Big Bang; that he further has stated that in the course of about a billion years, the primordial cloud formed by the explosion cooled from about 100 billion degrees to a point when it broke up into what are called galactic clouds or galaxies—billions of them; that in these galaxies, stars were formed out of the hydrogen and helium atoms, the first matter of the Universe and that subsequently over many billions of years, the present Universe was evolved, all without any guidance from any intelligent source or power but simply by continued accidental collisions of the hydrogen atoms in the star-formation period and later by the fusion of hydrogen nuclei in the hot cores of the stars where the heavier elements up to iron were formed. After iron, the heat source for element formation was provided by the fission of the nuclei wherein the breakdown of atoms resulted in extreme heat.

As we read the fascinating story which has been written by the men and women of science in just the last few decades, telling of almost everything that has happened since the first 100-thousandth of a second after the Big Bang, it becomes increasingly clear that the scientist has trouble with his own position and the dogmatic stand he has taken on the development of the Universe by pure chance. This is shown by the quite defensive statements made when it appears that his findings have built up an irrefutable position in favor of a guiding intelligence hovering over the Universe, bringing order out of chaos and setting up in the Universe the framework necessary for the kind of development that has taken place. He constantly refers to the "Laws of Nature"—the caps are mine—but is at a loss to explain how they were set in place.

As examples of the defensive attitudes expressed by people of the scientific community, Ditfurth writes: "A few observations seem appropriate at this point. We must not make the mistake of thinking that the sun developed these beneficial properties "in order that" life could develop on earth. Such a notion would presuppose that some external power had purposely arranged everything for our benefit. Actually just the opposite is true. The sun developed long before the earth and earthly life even existed. It certainly never tried to adjust itself to our future circumstances. Instead, life on earth adapted to the given conditions, which were largely determined by the properties of the sun. This fact does not make the development of life any less astonishing. All the same, some people tend to view the whole situation the wrong way around, betraying a narrow and anthropomorphic point of view" (*Children*, p. 104).

This is a revealing statement. It is evident that his own story about the development of the Planet Earth and the life on it has almost convinced him that there must have been "some external power" involved. He is checking his own feelings as much as those of his readers! Let us turn his statement around. If the Creator-God had willed to develop life on a planet, He would have prepared for it long before the existence of life on that planet. He would have provided a sun, such as ours; He would have built in the kinds of "life protection" systems that we have, to protect from too much heat, too much cold, too much radiation, etc. He would have provided everything necessary to sustain life. A solar system, such as we have, would have been a necessity. We will develop these thoughts more completely when we review the story of the galaxies, solar systems and life-formation as told by these same scientists.

Gribbin in *Genesis* (p. 307) shows his defensiveness with an even stronger statement. "It is equally true, though, that we are creatures of the Universe in which we live, evolved and adapted to fit our surroundings on a cosmic scale, although this may seem much less obvious at first sight. Indeed, some people find it so unobvious that the suitability of our Universe for life as we know it is sometimes raised as a serious philosophic or theological puzzle. Isn't it remarkable, runs this kind of argument, that the Sun is just the right temperature to keep the Earth warm, and that it stays just the right temperature long enough for life to evolve? Isn't it remarkable, indeed, that all the processes since the Big Bang have been just right to produce us—the way galaxies and stars are formed, the way planets formed, even such fundamentals as the strength of the gravitational force, seem to have been tailored precisely to the needs of people on Earth! In fact, of course, what has happened is that *we* have been tailored to fit our surroundings, including the strength of the gravitational force and the temperature of the Sun. Our form of life depends, in delicate and subtle ways, on several apparent "coincidences" in the fundamental laws of nature which make the Universe tick." The obvious, which evidently is completely unobvious to the scientist, is that his scheme of things will not function, or "tick," without the coincidences and the laws of nature. Gribbin goes on from this point to outline a number of "coincidences" without which life as we know it would not be possible. As with Ditfurth's defensive statement, the whole position should be looked at from a different perspective. The Universe as we know it and life as we know it are the results off the sustained, expressed will of the

Creator-God who put into effect the Laws of Nature before the Big Bang happened and who has guided the whole development down to the present. The concept of the scientific method breaks down completely when all the evidence is thrown out and the case for the origin of the Universe is based simply on the happenstance of accidental and random collisions of atoms in space.

A final witness at this point to the stand of the scientific evolutionist is taken from the writings of Paul Davies, professor of theoretical physics at the University of Newcastle-upon-Tyne, England, in *The Edge of Infinity*: "There is no apparent need for a supernatural organizer—the laws of nature themselves seem capable of generating the present high degree of structure and organization that makes the Universe so interesting" (p. 170). So, he needs the "laws of nature" but does not want the One who set these laws in place!

Thus, while the scientist has staked out his turf and appears to be adamant in holding to it, there appear to be some who are not very comfortable with what their findings indicate. So, as we pursue our studies of what the scientific community has provided us in the way of knowledge about our Universe, we will insist that the scientists be held accountable. This means that as we go on from here in reviewing their accounts of the Universe and of life, we will bear in mind that they believe that it all just happened to happen, "willy-nilly," to use Gribbin's phrase. It will become quite apparent that there is a purpose to what has happened and that is continuing to happen. This purpose is centered in a Divine Plan that transcends the thoughts of man, even as the Creator-God is greater than created stuff and living matter.

CHAPTER III
THE TURF OF THE WRITER

In order that readers may understand exactly where the writer stands with respect to his view of the Universe in which we live, it is necessary that he define his turf, the ground on which he stands. In doing this, it will be necessary to make plain a number of position points.

For the data on the Universe we will rely on the recent writings of the astronomers, astrophysicists and other scientists who have been doing such a remarkable job in these past few decades. It is fair to state that more has been learned about the Universe, its origin and development, in the period since World War II than in all the centuries before. Many of the laws of physics concerning matter, space and the heavenly bodies, laid down by scientists a century and more ago, have been confirmed. New understandings have also been developed and entirely new concepts have emerged, particularly by way of quantum mechanics, which have provided a much more adequate view of matter in the Universe in which we live. These findings have led the scientists into the submicroscopic realm of particles-waves wherein matter is formed and the beginnings of the Universe are to be found.

There has been an exciting breakthrough in two directions: the macrocosmic (the study of the Universe as we see it by way of highly sophisticated astronomical instruments), and the microcosmic (the study of the Universe as we can only learn of it by way of particle accelerators and other exotic pieces of equipment). All of this has come with the development of electronic brains, computors and related pieces that have powers beyond comprehension.

Along with these advances in technological capacities has come the breakthrough of travel into the space of our Solar System which enables scientists to make astronomical observations from beyond the atmosphere of the Planet Earth. Satellites are being launched, equipped with specialized instruments able to take radiotelescopic sightings and readings without the interference of the earth's atmosphere. In the near future, when manned space laboratories will be placed in orbit around the earth and when permanent bases will be set up on the moon, a new day will dawn for the gathering of data. Possibly the whole Universe will be opened to the scientists in their laboratories!

Thus, the latest published data by recognized scientific writers will be taken as being the most accurate. Developments in the field have been so rapid that information which was radically new just a few decades ago may well be obsolete now. Even as we write this, we recognize that by the time this book is published, items that appear now to be correct may no longer be so.

One thing must be clearly understood. There is no quarrel with scientists with respect to the discoveries that have been made and the new knowledge that has been gained in these last few decades about the Universe, its age or its make-up, whether macrocosmic or microcosmic. Nevertheless, some of the conclusions that have been drawn from the new knowledge may be open to question. Some of the positions taken as to the meaning of the data gathered need careful examination. It is not our purpose to contradict nor to deny the facts about our marvelous universe which the scientists have laid before us but rather to look objectively at the data provided. Out of this kind of study may come some possible alternative interpretations which could have the effect of strengthening the validity of the data while removing some of the anomalies.

The reader will not fail to notice the repeated use of the concept that "the obvious is often overlooked" as we seek to probe the significance of recent scientific discoveries. When a person reads backward in history, no matter what the field, this will likely be evident. Thus, my purpose is not to point a finger at the scientists engaged in the study of the Universe but to recognize that often those closest to a given situation fail to see in it some facet which to another person appears very significant.

Theologians also overlook the obvious. If there is something in the Bible that they do not understand or cannot explain, it has been easy for them to "spiritualize" or even ignore it. In our last chapter we quoted Jastrow, who acknowledged that this is also done by scientists when they find that their "beliefs are in conflict with the evidence." A case in point for the theologian to ponder is found in the Second Book of Kings, chapter 2, in which we are told that many people witnessed what we today would call a "soft landing" of a spaceship. With much noise, flashing of lights and smoke, the craft landed on the bank of the Jordan river near the city of Jericho. Elijah the prophet was taken aboard and the craft took off and was soon out of sight. The amazing thing about this incident is that for some time it had been known that this was to happen. They sought to prevent Elisha from continuing the journey. They apparently accepted the fact that Elijah was to be "taken up" but they did not want that to happen to Elisha. It should be noted that at that time electricity was unknown—flashing lights would be a mystery. What did the people think as they watched the event? The fact that this matter was openly known by many people and that they apparently were not alarmed about it or afraid for their well-being, forces the question to be asked: How often had they witnessed such an event? It is obvious that this was not the first time or there would have been terror in the hearts of the observors.

A search team of fifty soldiers were sent out to comb the nearby mountain sides in the hope of finding the craft in case it had crashed. That so vivid a description of a spacecraft landing could have been written at least three thousand years before our first modern spaceship was launched should be the cause of serious study. The theologians, however, either ignore it in their commentaries or they dismiss it with the statement that it was a symbolic event in which the people of Israel were to be assured that the "Chariot of Israel" would save them in time of trouble. The obvious that appears to have been overlooked is that a most remarkable incident had occurred. Hundreds of people knew of the appointment of Elijah to meet this craft at a specific place and time. They knew he would be taken up in a ship of some sort and that they would not see him again. Many people, possibly hundreds, actually saw the event take place. Such a happening needs an explanation if one can be found, or it needs an acknowledgement that this was something that goes beyond knowledge or understanding. There are other incidents described in the Old Testament which are similar to this one. All need to be studied in the light of the knowledge we now have of ships that can travel in space.

In this study, mention will be made at times of Bible passages, not implying that it is a scientific document but rather that it is used as a reference, much like the books written by the scientists have been used. One thing should be understood by both those who accept the Bible as the Word of God and those who don't. *Truth is absolute.* What is true with respect to the Universe as shown by the scientists will not be in conflict with what the Bible teaches about the Universe. There are areas in which some matters are not perfectly clear but these should not become a cause for disagreement between people who want to know the truth. These may well be the areas where careful study ought to be made by both the scientific and the religious, preferably working together.

The Universe is huge! To attempt to describe its size would be folly—except that it is in the very nature of man to attempt the impossible. There is no way for man not to do this for he has been given that kind of mind. Human beings are the only creatures that have a sense of time, of before and after. Man can, in a sense, stand apart from himself and study himself and his surroundings. He can study the past and learn of the times that have been and make plans for the future, the times that are yet to come. Man alone, in all the world, deals with *time*. This does not mean that he understands time. It is an enigma. We live in it, with it and by it. We can use it well to accomplish certain objectives and find that time passes very quickly. We use it poorly and feel frustrated, with time passing slowly. When we are bored, time almost stops—except that when we look back we discover that it has vanished and ask ourselves, "Where did the time go?"

G. J. Whitrow, reader in applied mathematics at the Imperial College of Science and Technology, London, in *The Nature of Time* states: "We now have abundant evidence that our sense of these temporal distinctions is one of the most important mental faculties distinguishing man from all other living creatures. It seems that all animals except man live in a continual present. Such examples as may be cited to the contrary do not survive critical investigation. Dogs frequently display powers of memory when they give vent to the wildest joy on seeing their masters after long separation; but this does not necessarily indicate any image of the past as such. Similarly, there is no firm evidence that any animals have a sense of the future. Carefully analyzed experiments have shown that, even in the case of the most intelligent animals such as the chimpanzees, any actions they take that might be thought to indicate some such sense are in fact purely instinctive. . . . In man's case, awareness of the distinctions between past, present and future must have been the result of conscious reflection on the human situation" (pp. 9–10).

We really do not know what *time* is. Before you were conceived, for you there was no time, no yesterday, no today or tomorrow. For you, TIME did not exist, for you did not! For our *Universe*, the same is true. Before the Universe came into being, there was no time relative to it. There must, however, have been a "present" in view of the fact that the Universe exploded into being. This present would thus have to be in a different time frame, such as the *eternal present*. If that be so, there must also have been a *before* relative to the newly-born Universe and it follows that there will also be an *after* relative to the end of the Universe. Could it be that the whole of Universe-Time is at once encompassed by the eternal present?

Paolo Maffei, in his book *Beyond the Moon*, takes his readers on an imaginary journey through the Galaxies and out into the far reaches of space to what he believes is the end of the Universe. "At this point we can truly say that we have arrived at the limits of time and space. We shall no longer be able to wonder at what lies before or beyond, for there is no longer any before and beyond does not exist" (p. 325). Maffei in his travels started from a point where he felt he could say that ahead lay a boundary that marked the end of time-space. He drew the conclusion that there could be nothing beyond that point. Was he right? Or, within the concept of the *eternal present* could it be said that the beginning of TIME as we know it—"universe time"—came with the Big Bang, out of the eternal present? The end of time as we know it could come in a return of universe time to the time frame of the eternal present. Neither the theologian nor the scientist has given enough attention to the concept of the eternal. The Bible clearly speaks of the eternal and of matters that go beyond our universe time, both before and after.

St. Augustine is reported to have said, "But what really is time? As long as no one asks what it is, I seem to understand it perfectly well. But if I am asked to *explain* time, I am suddenly struck dumb." Like Augustine, we all think we know about time. We have counted it, we have divided it into segments such as years, months, weeks, days, hours, minutes, and seconds, but this does not indicate that we know what time is. Rather, we know something about how to use it. This is very helpful—at least up to this time. There may well come a point in time when we will get a new view on it, a new dimension. Thus, in our present consideration, there is no problem with respect to the enormous amount of, or length of, time involved in the age of the Universe, the Galaxy, the Solar System or the Planet Earth. Time, for those who live in the Universe, is relative to the duration of the medium in which we live.

The existence of the Creator-God before the Big Bang is demanded by the logic of the situation. Without such an existence there would be no explanation for the Big Bang nor the existence of the Matter-Energy that exploded the Universe into being. The

Bible simply states that "In the beginning, God created . . . " The Hebrew verb *bara* means to "create ex nihilo." The account of the Bible presents the story of the beginning of the Universe and, following that, the specific development of the Planet Earth and the life on it. The story had been handed down by word of mouth from dim antiquity and was put into writing by an unknown hand relatively recently. Nowhere does the Bible provide a timetable with respect to the creation or the beginning of the Universe. Neither does it indicate when life, plant or animal appeared. It does, however, indicate an order of the appearance of light, plant life, animal life and human life. This order is identical with that which the scientists have determined in their studies. This is an assurance that what is found to be true in one field will not be in conflict with what is true in the other.

In the 17th century, Bishop Ussher prepared a chronology of the Bible, setting dates for important events. This chronology was placed in the margins of the English Authorized Version by the Bishop of Worcester. According to this chronology the creation of the earth occured about four thousand years before the birth of Christ. Thus the planet would be only about six thousand years old. Unfortunately, because Ussher's chronology was placed in the margins of that edition of the English Bible, many people came to believe that this age for the Planet is what the Bible taught. Great problems have been caused because of this error. Those who accepted it as fact found themselves at odds with all the scientific evidence on the age of the Earth and the Universe. Ironically, they also hold a position that is not in accord with the teachings of the Bible, for it clearly speaks of the Creator-God as is an eternal Being, not confined to a few thousand years. The ancient Psalmist wrote in Psalm 90, "Before the mountains were brought forth or ever thou hadst formed the earth and the world, even from everlasting to everlasting, thou are God . . . For a thousand years in they sight are but as yesterday when it is past, and as a watch in thy night."

These words simply mean that time, as we measure it, has no meaning with respect to the eternal Creator-God. The Church has never adequately addressed itself to the concept of time in relation to the Creator-God and the Universe which he made. Had it done so, much of the constant strife between those who contend that the earth was created recently with no apparent relationship to the Universe of which it is a tiny part, and those who hold with the astrophysicists that the Planet has been in existence for about four and a half billion years, could have been averted. The proponents of the recent-creation theory apparently do not want to admit to a relationship between this planet and the other bodies in the Solar System, to say nothing of the relationship of the Solar System to the Galaxy or the Galaxy to the Universe.

From the Bible we learn that everything given to man was good. It was not "in the will of the Creator-God" that man should know evil. Nevertheless, in order that the human could be a person in His image, it was necessary that he be given the opportunity to freely choose to live in obedience to the will of the Creator-God. This marks a basic difference between the human and the animal, the ability to reflect and exercise judgement in accord with his free will rather than to be governed by instinct with no capacity to conceptualize or make judgments reflecting opinion as to what is good or evil. This means that with the gift of free will there was also a corresponding responsibility for one's decisions. Again, this makes a true distinction between the human and the animal, the first being responsible for his acts and decisions whereas the latter, ruled by basic instincts, would not be held responsible. The gift of freedom of the will carried with it the possibility of an eternal fellowship with the Creator-God on the basis of the free choice made by man. It also carried with it the possibility of a breaking of fellowship, the consequence of choosing disobedience. This was the course man chose and the results are plainly evident throughout the history of man's life on earth. God knows evil in more terrible ways than any human can comprehend. He knows *absolute evil*. We will touch on this again.

It is difficult for many people to accept the concept of a transcendental Being who has existed forever and is responsible for the creation and the sustaining of the Universe. The scientist, and many others, who struggle with the agnostic problem, are saying in effect, "Unless I can reduce the Being of God to something I can handle, something small enough so that my mind can comprehend it, I will not believe." Thus, God must be reduced, not only to the level of man but to something lower than man, for this reasoning insists that God must be scrutable, understandable, explainable. This is an absurdity! God must be inscrutable, beyond comprehension, above and beyond all that is in the experience of His creatures. This is why Moses was told on Mt. Sinai, "Thou canst not see my face for there shall no man see me and live" (Exodus 33:20). Jastrow in *God and the Astronomers* puts it like this: "When an astronomer writes about God, his colleagues assume he is over the hill or going bonkers. In my case it should be understood from the start that I am an agnostic in religious matters" (p. 11). This position creates more problems than it solves for the agnostic refuses to take a stand on the basis of evidence at hand which explains the otherwise unexplainable. In doing so he actually takes a stand on the basis of accepting postulates stemming from the limited capacities of the human mind and experience which really denies what under other circumstances would be accepted as empirical evidence.

The scientist has a special problem here, for being committed to the scientific method, examining the evidence for or against a proposed hypothesis, he discovers that dealing with a transcendental Creator-God goes beyond the limits of the laboratory. He therefore turns from taking a step in faith and chooses to stand on what he calls "scientific ground." In this manner he rules out the possibility of a Creator-God. At the same time, however, he fails to understand that his option—that the Universe came into being by happenstance—has no scientific basis. Thus, he also acts with a kind of faith and builds his Universe-concept on what he has deduced about the Big Bang of Creation, ignoring the fact that the Source of it all has not been accounted for.

The Creator-God has revealed himself with remarkable clarity in providing the Universe, in establishing and sustaining the Laws of Nature which govern it, in the creation and development of life in the many forms by which we know it on this planet, and in providing man with the intellectual capacity to explore, to learn the secrets of the Universe, the Laws of Nature and the ability to use them. To those who accept the Bible as His word, He has further revealed Himself in the message of that book.

This book is "science friendly" in its orientation. It has as a goal, the elimination of the long conflict between science and religion. There are some things, however, that we must face when we deal with the scientist and his position with respect to the existence of the Creator-God. We do not mean to imply that all scientists are atheistic. Many are believers in the Creator-God, some are agnostic, and some just don't know where they stand. While we have respect for them and the truly magnificent work they have done in tracing the beginning of the Universe back to within one hundred thousandth of a second after the Big Bang, we do not stand in awe of them! Instead, we remind them that the scientific community is unable to prove that the Creator-God does not exist. As the reader will notice, we have often cited instances where the scientist in the process of denying the existence of a Super Intelligence, resorts to using personal opinions, even fantasies, in seeking to explain away the "coincidences" which he has had to insert in his account in order to have a viable Universe. The evidence for the existence of the Creator-God that has been abundantly supplied by the scientists has caused many of them to take space in their writings to warn their readers not to entertain any foolish anthropomorphic ideas about the possibility of the existence of such a Being. The scientific community has taken the position that it can go only so far in describing the history of the Universe. They always stop short of stating that on the other side of the Big Bang is the First Cause, the One who is responsible for the creation of the Universe—the Creator-God! Perhaps the scientist is right—he can't go beyond that point until he is ready! Maybe the theologian is the one who should push on! I realize that the scientists might well ask the theologian to prove that the First Cause exists. No problem! That is what the believer has always known. He does not have to "prove" it. It is not in the rules that the scientist is to explain anything that goes beyond the scope of the Universe as we know it. The theologian, however, is told specifically in the Bible that he cannot see the living Creator-God, for God is inscrutable and His ways past finding out. So, the scientist should not feel hurt when told that the final answer has not yet been revealed. He and the theologian should join in describing the wonders of the Universe which the Creator-God has made.

Another thing of great significance in our study is that the development of the Universe, from the moment of the Big Bang to the present, has been through a process that can best be described by the word "evolution." This may surprise some who think of evolution as something that happens only to plants and animals, organic life. This is not so; the entire "growth" of inorganic particles-waves into elements and molecules up to the present galactic systems, stars and all they contain, including the houses we live in and the cars that take us to work, all has come to its present form through the process of evolution. Not the "willy-nilly" kind of accidental happenstances of the scientific evolutionists but under the rule of the Laws of Nature set in place by the Creator-God at the time of the Big Bang of Creation. Underlying these laws was the basic equation that in a real sense describes the relation between the Creator-God and all of Creation: $E = MC^2$. This concept we will discuss in several places as we move on through the development (evolution) of the Universe as it changed from being pure energy into being matter out of which the Universe as we know it came into being. When we recognize that it was less than one hundred years ago that Albert Einstein brought this equation to the attention of the scientists of the world, we realize with a shock that relative to universe time it was just a tiny part of a moment ago that mankind began to understand and use this equation.

In the meantime, during that very long period of universe time, almost the whole of it, less one hundred years (!), the Creator-God and His workman (they will be identified later on) were evolving the Universe as we have come to know it. They are still at work! Our Solar System with its Sun and planets, are very recent arrivals on the scene. Most astrophysicists agree that the Big Bang exploded about 15–18 billion years ago. Our Solar System is a young 4½ billion years! But when the time came for our Solar system to form, there also was the presence of the Creator-God. We read in Genesis 1:2, "And the earth was without form, and void, and darkness was upon the face of the deep. And the Spirit of God moved upon the face of the waters. And God said, Let there be light, and there was light."

This was not the light of the Sun as we know it but the pale yellow light of a Sun just being born! Later, when the Sun had grown strong and nuclear fires burned in it, the light broke through and the solar wind began to blow, clearing out the dust and debris of the space of the Solar System and doing many other wonderful things of which we will learn later. What we need to recognize is that here, at that time, the Creator-God was present. He never was absent! He will continue to be present until His purpose has been fulfilled and universe time with all of the

Universe in its hugeness will be brought back into the eternal present in a manner we cannot comprehend now.

We have mentioned the Laws of Nature and the basic equation that undergrids all of the Universe. In order to be functioning, however, the Laws must have something on which to work. There has to be matter-in-motion! The Creator-God provided for this in the Four Forces of Nature that came into being in the explosion of creation. The strong force dealt with the formation of protons and the "holding together" of matter, the electromagnetic force dealt with all things electrically, and the weak force dealt with the radioactive decay of matter. Each of these forces work as though commissioned to do these several things and no more. The fourth force was gravity which deals with everything and yet is not affected by the actions of the other forces. It works through the capacity of being "attractive" to all matter in the Universe. We will deal again with this force which does not answer to the rules of quantum mechanics but which exerts force over all the Universe. The other "forces" appear to have "come out of" the force of gravity at the time of the Big Bang, without in any way diminishing it. The scientific community does not yet know just how to describe its relation to the matter of the Universe in the same degree as is the case with the other forces.

The strong force binds the quarks together as hadrons on the way to becoming protons, the nucleus of atoms. The electromagnetic force provides the positive electrical charge for the proton. It works through the leptons to provide the negative charge to the electrons. The weak force enables radioactive elements to decay in the process of element formation and in doing so provides energy for other reactions. Whatever happens in the Universe appears in some way to be under the influence of the fourth force, the power of gravity. While appearing to be the weakest of all the forces, it is really the one that rules them all! At this point we would say that the force of gravity is the control agent of the Creator-God, always present establishing order in the Universe, guiding the evolution of matter into becoming what is in the will of the Creator-God for this Universe.

The chief goal of the evolutionary process in this "before-life" period was the development of the "life-molecule" a form which would, when powered with the essence of life, become the first living thing—a microscopic plant form with a photosynthetic process in place so that power from the rays of the Sun could be translated into energy for the life of the plant and its self-replication. The primary life-form evolved into higher forms which, as time passed, produced all life forms, even the animal and human.

So, while Charles Darwin is given credit for providing a description of the evolutionary process for the development of plant and animal life as we know it, the process actually began at the time of the Big Bang with the formation of the first matter forms—particles-waves—which became elements and led to the physical Universe we now know. Thus, the scientific community of yesterday and today must be given credit for having described the evolution of matter from the beginning of universe time to the present, and the process goes on from now to when the will of the Creator-God will have been satisfied. The high-tech knowledge, the sophisticated instruments, the particle-accelerators, the capacity to fly people to the Moon and return them safely to Earth, are just breakthrough items. The scientist, aided, we hope, by the theologian, will produce wonders in the next century so great we can't at this time comprehend what all will happen.

Von Ditfurth states: "He laws and forces ruling the Universe justify our calling it a 'cosmos', a harmonious entity whose every part is responsive to every other. . . . The depths of space are ruled by the same laws and forces which govern terrestrial life . . . a planet did not come into being independently of the rest of the Universe" (*Children*, p. 12).

The obvious, which he and others with him, have overlooked is plain: while all of the truly magnificent achievements in astronomical science which trace the development of the Universe back almost to the beginning moment, demands that there must be a "Beginner," a "First Cause," the scientific community stubbornly refuses to admit it. Jastrow, in *God and the Astronomers*, in attempting to deal with the dilemma of the astronomers on this point states on page 103, 'There is a kind of religion in science; it is the religion of a person who believes there is order and harmony in the Universe. Every event can be explained in a rational way as the product of some previous event; every effect must have its cause: there is no First Cause."

The evidence is in, however. The Universe did have a beginning and therefore a First Cause who brought it into being in a cataclysmic explosion or creative act from which a faint radiance is still bathing all of space. The development of the Universe, which we will soon be looking at, was in good hands, for He who created it had a purpose and with that purpose there were goals and with them, responsibilities. Laws of Nature were set in place and have been sustained. In ways which we may not yet understand, order, and in time, life, in all of its varied forms were brought out of the original chaos as the Spirit of God brooded over His Creation.

CHAPTER IV
THE BEGINNING OF THE UNIVERSE

The view of the modern scientific community about the beginning of the Universe has firmed up to substantial agreement in this century. This view was actually predicted when Albert Einstein presented his theories on General and Special Relativity, even though it was not recognized at first. It is interesting to note that while his theories demanded an expanding universe, Einstein refused for a time to admit the fact. He was committed to the "static state" concept, that the Universe had always existed and would continue to exist forever. To admit that the Universe had always existed and would continue to exist forever. To admit that the Universe was expanding meant that as time was viewed backward, the Universe would get smaller and if one looked far enough back in time, it would come to what scientists call "a point of singularity," meaning a beginning point where all that is in the Universe would be compacted with such extreme density that no matter existed, only radiation in which subatomic particles were in constant violent interaction. Jastrow states that all the Energy-Matter of the Universe was compacted in the nucleus of an atom! This thermodynamic equilibrium could not last long; it had to explode. Jastrow writes, "At that time all the matter in the Universe was packed into a dense mass, at temperatures of many trillions of degrees. The dazzling brilliance of the radiation in this dense, hot Universe must have been beyond description. The picture suggests the explosion of a cosmic hydrogen bomb. The instant in which the cosmic bomb exploded marked the birth of the Universe" (*Astronomers*, p. 13).

If the Universe had a beginning, the next question would be, "How did it start?" It was likely this point that Einstein did not want to consider. A Dutch astronomer, Willem de Sitter, was the first to discover that the equations of Einstein predicted an expanding Universe. Later, a Russian mathematician, Alexander Friedmann, made the same discovery. To de Sitter Einstein wrote, "This circumstance [of an expanding Universe] irritates me," and in another letter about the expanding Universe, he said, "To admit of such possibilities seems senseless." He ignored a letter from Friedmann which described his solution to the equations. "When Friedmann published his results in the *Zeitschrift für Physik*, in 1922, Einstein wrote a short note to the Zeitshcrift calling Friedmann's result "suspicious' and proving that Friedmann was wrong. In fact, Einstein's proof was wrong." Later, in 1923, Einstein acknowledged his error, writing to the *Zeitschift*, "my objection [to the Friedmann letter] rested on an error in calculation. I consider Mr. Freidmann's results to be correct and illuminating."

Many scientists have felt just as uncomfortable about the situation as did Einstein. "In fact, some prominent scientists began to feel the same irritation over the expanding Universe that Einstein had expressed earlier. Eddington wrote in 1931, "I have no axe to grind in this discussion, but the notion of a beginning is repugnant to me. ... I simply do not believe that the present order of things started off with a bang. ... The expanding Universe is preposterous ... incredible ... it leaves me cold" (Jastrow, *Astronomers*, pp. 112–113). The German chemist, Walter Nernst, wrote, "To deny the infinite duration of time would be to betray the very foundation of science." More recently, Phillip Morrison of MIT said in a BBC film on cosmology, "I find it hard to accept the Big Bang theory; *I would like to reject it*." And Allan Sandage, of Palomar Observatory, who established the uniformity of the expansion of the Universe out to nearly ten billion light years, said, "It is such a strange conclusion ... *it cannot really be true*." (The italics are Jastrow's.)

Even today, it would be fair to state that some scientists would like to find another way to resolve the problem. Failing to find such a solution, they appear at times to simply ignore the problem. "Picture a world of pure energy flashing into being, light of unimaginable brilliance fills the Universe, the cosmic fireball expands and cools; after a few minutes, the first particles of matter appear, like droplets of liquid metal condensing in a furnace" (Jastrow, *Sun*, p. 21). The Universe, according to his account, simply happened to explode into being! No explanation is given of the meaning of "pure energy" or of how it came to be.

Paoli Maffei, in *Beyond the Moon*, wrote of that beginning, "At that time all the matter and energy in the Universe must have been concentrated at the enormous temperature of 100 billion degrees, in a sphere that the astrophysicists have named the fireball. [Note the vast difference in temperature between Maffei's estimate and that of Jastrow just quoted, a temperature in the trillions of degrees.] This sphere had a radius scarcely one ten-billionth that of the present universe and was composed of protons and neutrons, radiation of extremely high density, and a gas of photons, electrons and neutrinos—the magic moment of the origin! It is calculated that in only one second the temperature of the fireball would have fallen from 100 billion to 10 billion degrees, while the radius increased a good ten times. After scarcely a hundred seconds the radius would have increased a hundred times. This is not just an expansion, but a true explosion of the universe—an inconceivable explosion. From it the Universe was born, and its dizzying course is still continuing today; we do not know when, or if, it will ever stop" (p. 322). Again, we are told that the Universe came into being in a "magic moment" with no apparent cause.   John Gribbin, in *Genesis*, states, "This, in a nutshell, is the picture which tells us that there must have been a Big Bang. First, just by imagining the Universe "running backward" we can see that it has evolved from a state of greater density with clusters of galaxies closer together; go back long enough and everything must have been squeezed in one lump. And, secondly, the Einstein/Friedmann models which so successfully describe the state of the Universe today all begin from a Big Bang, an initial birth of creation when space-time/matter-energy burst out from a unique point, a mathematical singularity" (p. 23).

Implicit in this statement is the assumption that there was no Creator-God, who from his "non-universe" existence called into being the fireball and all that was in it. The scientist refuses to see the lack of correlation between his theory of the Conservation of Energy and his position that the Universe just "happened" into being. He also does not appear to realize that when he describes a big bang as being "an initial instant birth of creation when space-time/matter-energy burst out from a unique point, a mathematical singularity," he is in effect saying that there must have been something "before" the Big Bang, located at his "unique point."

Von Ditfurth states: "In the beginning was hydrogen. Besides hydrogen, nothing else existed but the laws of nature and space itself—unimaginable quantities of space. The world was born from an immense cloud of hydyrogen gas which began to contract under the influence of its own gravitation. The process of contraction began some 10 to 15 billion years ago. Our entire Milky Way Galaxy, including the earth and ourselves, developed from this primal cloud" (*Children*, p. 283). He was writing of the cloud of energy particles and gases that resulted from the fireball explosion. Von Ditfurth apparently chose to take the moment after the explosion of creation as the beginning point for the Universe and thus ducks the basic question as to the origin of the mass of hydrogen out of which he says the Universe developed.

These views of eminent scientists, published within the last few decades, fairly represent the position of today's astrophysicists with respect to the beginning of the Universe. It was born in a cataclysmic explosion so huge, so powerful, that it really is folly to attempt to describe it. All of the Energy-Matter of the present Universe was contained in the fireball that exploded. The story of what has happened since then has been told quite convincingly back to the first part of a second!

Because it had a beginning, most scientists believe it will have an end. "The lingering decline predicted by the astronomers for the end of the world differs from the explosive conditions they have calculated for its birth, but the impact is the same: modern science denies an eternal existence to the Universe, either in the past or in the future" (Jastrow, *Sun*, p. 30). The scientists may well be correct with their predictions about the Universe as we know it, but they have not defined its beginning!

With all respect due these scientists for the magnificent work they have done, there is a place for a second opinion. In keeping with the Einstein formula, $E = MC^2$, energy and matter are interchangable, but none of it can be lost. By the same token, none can be created in the Universe as we know it. Thus, there must have been a "Before" in an "Out-of-this-Universe" situation where the Creator-God not only brought into being the subatomic energy particles of our Universe but set in place the Laws of Nature which have governed all that has happened since the Big Bang. As we will see, this is consistent with all that the scientists have discovered.

The scientific community has done an excellent job in describing the beginnings of the Universe and its subsequent development down to the present. It is amazing to see how the bits of information gathered by many people over a long period of time, and the enormous quantities of information which have been assembled in the past few decades, fit like the pieces of a jigsaw puzzle to give us a very good description of all that has happened from the time of the Big Bang of creation to the present period when man, having discovered some of the mysteries of the atom and of its awe-full power, appears to be on the threshold of blowing up the little speck in the Universe that we call Planet Earth.

One problem, however, that appears to be misunderstood or possibly ignored by scientists is that description is not the same as explanation! No matter how well the process of the development of the Universe has been described, the nagging question of *how* it all happened remains unexplained. The concept that the Universe came into being by accident, without a cause, in a violent explosion which resulted in an enormous cloud of radiation out of which were developed, without the guidance of any intelligence, all of the particles of Energy-Matter that later on would become hydrogen atoms, is the basis for the scientists' view of the formation of the Universe. This concept is totally contrary to the absolute order we observe in the functioning of the elements of the Universe. If the development of the Universe after the Big Bang, which happened by accident, is also by happenstance, it follows that the substance of whatever it was that exploded must also have been assembled willy-nilly!

The scientist has been able to describe in minute detail all that has happened since the first one-hundred-thousandth of a second from the time of the Big Bang but has not been able to come up with an answer to the question of how it happened. Now that the nucleus of the atom, the proton, has been broken down into many parts, it is incumbent upon the scientist to come up with a scenario for the pre-Big Bang assembly of the parts of the proton. And again, it must follow the concept of randomness! Protons are not self-replicating entities. Each one must have come into being independently. Each one can be split into thirty-two fundamental particles. The laws of probability would be stretched to the breaking point to allow for so many accidental happenings in the creation of the almost infinite number of protons that exist in the Universe. And that is just a part of the problem! The existence of all of the electrons, neutrons, photons and neutrinos must also be explained, to describe their actions or functions is not sufficient.

The original cloud from the Big Bang cooled rapidly as it expanded at great speed. It also broke up into billions of huge blobs of gas—75% hydrogen and 25% helium. The formation of the first atoms of matter had begun as soon as the temperature dropped to a point that would allow the extremely hot subatomic energy particles of electrons, protons and neutrons, to enter into stable combinations. Until this happened, all the substance of the Universe was in the form of hot radiation in which energy particles or waves formed and disintegrated continuously. In the gas cloud there were also the photons (light waves) and the mysterious neutrinos which carry almost no weight but travel at the speed of light. They are so tiny they can pass freely through what we call "solids," even lead.

The huge clouds that formed as the original cloud broke up became what we call the Galaxies. The stars were formed out of the hydrogen gas. This we will discuss in our next chapter. In the meantime, we must look back to see what happened at Time Zero, when it all began.

In order that we might begin this exploration with some hope of keeping it coherent and with a determination to bring into the description the proper explanation of how it happened, it will be necessary to set some parameters, the first being that this chapter will deal only with what happened during what we might call the first general period, the time of cooling to the point where Galaxy formation began. This was roughly a period of about one billion years!

We need also to take a look at the size of the Universe. Here we will rely on the findings of the astronomers who tell us that the Big Bang of creation occurred about 15–20 billion years ago. As there are differences of opinion on this matter, I will choose to go with those who say that as the rate of expansion is gradually slowing down due to the tug of the gravitational force, the age of the Universe should be set at 17.5 billion years, or midway between the first two figures. This has been determined by the very sophisticated use of astronomic equipment, enhanced by the use of advanced computers in the hands of very knowledgable scientists. We do not question their figures. The results, however, are actually incomprehensible to our minds for we really have no words for the numbers that must be used.

Astronomical distance is described in light years—the distance light travels in one second times the number off seconds in a year (about six trillion miles)—and light travels at about 186,200 miles per second! To find the size of the Universe, we must multiply the numbers of miles light travels in one second by the number of seconds in 17.5 billion years! We use the term "size" of the Universe because we believe that most astronomers subscribe to the theory that the Universe had a beginning and that it will have an end. Therefore, it must have some dimensions. We will try to determine the distance to its outer point from our high point of observation by multiplying 186,200 by 60 seconds, by 60 minutes by 365.25 days by 17.5 billion years. This would give us the number of miles a hypothetical arrow would travel from our high point in that time. As it is believed that the Universe is spherical and that it is isometric and that the Cosmological Principal is correct (that on a large scale the Universe is everywhere the same), it would follow that the hypothetical arrow shot out from our high point would, after going the distance, end up by hitting us from behind! The distance travelled would be about 95,587,611, 880,000,000,000,000,000 miles—which would be the circumference of the Universe! So, the Universe is huge. Our planet, and even our Solar System is in reality just a speck, too small to be located in the enormous volume of space. This is the macrocosmic view of the Universe, a look at its hugeness. We will look at this aspect of the Universe more in detail when we study the formation of the Galactic System.

Having thus "looked" at the Universe at large, we must also understand how small the parts of it are. Indeed, without knowledge of the tiny components which make up the mass of the Universe, we could never come to an understanding of it. This involves the microcosmic view. In that study, we will learn some astounding facts, which make some scientists uneasy for they just might be forced to take a new view of the Universe, a view that would discover design and purpose in all that has happened. The existence of the Creator-God would then have to be faced.

The Universe is made up of 103 natural elements. The atom is the basic unit of an element and for centuries it was believed that it was the smallest unit of matter and not divisible. Now we know that while it may be the smallest unit of true matter, it has a complicated structure which can be broken down. Hydrogen is the lightest element and the first to come into existence. It has one electron in orbit around one proton. This atom appears to be the basic building block for all the other elements. When we discuss the make-up of atoms of elements, we will look very closely at the hydrogen atom for all of the matter in the Universe appears to be built up out of it.

At this point, in setting some parameters, we are only interested in indicating how very small the atom is. If you were to have one cubic centimeter of hydrogen atoms in a tiny box about one-half inch square, you would, if you were at sea level, have 25,000,000,000,000,000—or twenty five quintillion atoms in the box! Another way of measuring the size of an atom would be to imagine that 250,000,000 atoms were laid end to end. The length of the line would be less than an inch. If 100,000 electrons, an essential part of an atom, were laid side by side, they would just cover the width of one hydrogen atom. Still another game to play: if the proton, the other essential part of the hydrogen atom, were to be as large as a golf ball, the electron in orbit around the proton would be one mile distant from it. Thus, an atom is mostly empty space! The proton, while exceedingly small, provides the mass (weight) for the atom and determines its nature.

Another energy particle, also found in the original fireball cloud, is called a neutron. It has no electrical charge, as its name would imply. The neutron has the property of entering the nucleus of atoms. It does not change the nature of the atom in doing so but it does increase its weight for it carries the same weight as a proton. It can slip into the nucleus of an atom without increasing the atom's size. Thus, we have seen that the Universe is extremely large but not infinite; it has dimensions even though they are so large we cannot really comprehend them. It likely is in the form of a sphere for the force of the Big Bang moved out in all directions with uniform power. It had a beginning and it will have an end.

At the same time, we have looked at its constituent parts, the particles which were in the Big Bang Cloud of creation and find that they are exceedingly small. The tiny nucleus of the hydrogen atom, the proton, small as it is, has been split into thirty-two still smaller parts! There is no way to describe them except to state that they are energy forms or waves—non-material—and that they have specific functions in the make up of the proton. The electron is a versatile energy wave as it moves with incredible speed in orbit around the proton. It is believed to be a "fundamental" particle, meaning that it has not been divided. The electron has many functions. It is involved in the structure of all elements, even as the proton is, but beyond that it plays a significant role in the bonding of elements to form molecules of matter. When we discuss at a later point, the cells of plant and animal life, we will also see that it is an absolutely necessary source of energy in translating the energy of the Sun to the plant and of the plant to the animal cell. In addition, electrons in motion through a conductor constitute electricity! The electron has a negative electrical charge, whereas the proton has a positive charge. Thus they are attracted to each other, but a "force" which we cannot explain here prevents the electron from crashing into the proton. The expressed sustained will of the Creator-God enters into this matter. If that did not happen, the crash of the electron and the proton would make the creation of elements of matter impossible—there could not be a Universe! These two energy forms out of which all the material stuff of the Universe is made are not of themselves "material." What then are they? We will suggest an answer a bit later.

It is utterly impossible to conceptualize the situation at the time of the fireball explosion. The almost infinitely dense and hot mass of energy (no atoms of matter existed) was being hurled out into space with such a force that the explosion of innumerable hydrogen bombs would be too weak a comparison. "The dazzling brilliance of the radiation in this dense hot Universe must have been beyond description. The picture suggests the explosion of a cosmic hydrogen bomb" (Jastrow, *Astronomers*, p. 13). Electrons and protons swirled in what, for lack of any other word, we must call a cloud. In that cloud, which was expanding at an enormous rate, were neutrons, energy particles which play an important part in the development of elements. As we have mentioned, neutrons have the peculiar function of making atoms of the elements heavier without changing the nature of the element. A hydrogen atom with one neutron in its nucleus is called "heavy hydrogen" and has the name deuterium. A deuterium is an "isotope" of hydrogen, meaning that it still has the same nature as hydrogen even though its weight has doubled. All atoms, of whatever element, can take on neutrons and thus have isotopes. Some have a number of them, tin having the most at ten. This property is of great significance in the building of both elements and molecules, as we shall see later.

There were also the waves or particles called photons in the radiation cloud after the Big Bang. They carry no electrical charge and are practically weightless. They are light waves. The fifth energy particle or wave to be found in the creation cloud was the tiny neutrino which appears to be just a burst of energy that travels with the speed of light. They have almost no weight. It is not understood with respect to function but is nevertheless considered to be of great significance.

In addition to these "real" particles, it is believed that there were also counterpart waves or particles of "anti-matter." It is believed that when "real" particles collide with "anti-matter" particles, both are annihilated, releasing a great burst of energy. The scientific community has not really defined what is meant by "pure energy." We will seek to do that later on in this chapter.

An explosion of the magnitude of the Big Bang must have produced sound waves of fearful intensity. But all had been quiet, no sound was heard, for there were no ears to hear, no receptors to record. For us who are accustomed to hearing even the sound of the cricket, it is incomprehensible that in the billions of years that followed the creation of the Universe, all was deathly silent in spite of the repeated explosions in the dense clouds of Energy-Matter which were spreading rapidly out into the space of the Universe. The light that flashed with a brilliance that would have blinded the eyes of any beholder, was not seen. Only in our minds can we now try to visualize what it was like when the Universe was born. We must not confuse the light of the early period with light from a Sun, for there were no suns at that time. It was "pure" light composed of the photon particles-waves which were in the fireball cloud.

The basic building blocks of matter as we know it in the Universe were the electrons, protons and neutrons. These would, when the cloud had cooled sufficiently, form the first natural element of matter (the hydrogen atom) in which one electron would go into sustained orbit around a specific proton nucleus. As this happened, almost the whole fireball cloud became a mass of hydrogen gas out of which all the present elements of matter in the Universe were to come. It was likely for this reason that Hoimar von Ditfurth began his story at this point, when he said, "In the beginning was hydrogen."

At the moment of the explosion all of that which was created consisted of intense radiation at extremely high temperature and density. It was too hot for electrons and protons to form stable atoms. "So one-hundred-thousandth of a second after the beginning, the Universe was a seething mass of particles and radiation, a swirling soup in which pairs of particles were constantly being created from radiation, and constantly being destroyed and turned back into radiation" (Gribbin, *Genesis*, p. 33). Again, we have the scientist using a term without defining it. Was it pure energy? If so, we need to define that. This we will try to do (on a non-scientific basis) a bit later. At this point, the total mass of the Universe was kept constant, for as many particles as were formed by radiation were also destroyed by the collisions of "real" particles with their counterparts, the "anti-matter" particles.

Due to the speed of expansion, the Universe cooled very rapidly—the same amount of heat being spread more thinly in the increasingly larger volume of space which, according to Gribbin, was being created along with all else. At the end of the first one-tenth of a second the circumference of the Universe would have been about four light years, or 24 trillion miles! If this appears to conflict with Einstein's equations which indicate that nothing can travel faster than the speed of light, it does! There are possible answers to the problem. Gribbin states that the stuff of the Universe at that time was "carried along for the ride" by the expanding sphere of the Universe and thus was not in itself "moving". It would appear to this observer that a better solution to the problem would be to recognize that Einstein's equations refer to the travel of matter. At this point in time, there was no "matter." All was radiation, pure energy, and thus "non-material" and not subject to the speed-of-light limitation which Einstein's theories place on matter. In support of my point, I would use the statement of Carl Sagan in *Cosmos*: "No material object may move faster than light" (p. 200). This, too, we will touch on later in this chapter as we think of the meaning of the "non-material" substance of the fireball cloud.

"About fourteen seconds after the Big Bang", Gribbin writes in *Genesis*, "the temperature of the Universe had dropped to around 3×10^9 (3 billion) degrees and the weakening radiation no longer had the strength to create pairs of electrons and positrons. Most of the electrons met up with their opposite numbers and annihilated; the great days of mass/energy exchange were over, and the Universe became a lot quieter and a lot emptier" (p. 35). Descriptive language can sometimes cause problems: the "great days" in the first fourteen seconds? No sound could be heard for there were no ears to hear and the Universe was not "emptier" for the same mass/energy was present. It just was spread out more thinly in space.

The original energy-mass-radiation must still have been present in the Universe but in different forms. As the cooling process continued, the period of element-formation began. Nuclei of hydrogen and helium were formed. "By the end of the first four minutes after the Big Bang, 75% of the remaining mass of the Universe was in the form of hydrogen nuclei and the rest had been processed into helium nuclei; it took another 700,000 years for cooling to proceed to the point where electrons became bound to the nuclei to make atoms at a temperature around 5,000K." Thus, after so many years the matter of the Universe became stable in the form of the two earliest elements, atoms of hydrogen and helium. It is at this point that we can begin to tell the story of the development of the Universe as we know it today; the formation of the Galaxies, the stars, the Planets and finally life on a planet we call Earth!

In order that we might understand as well as possible the fact that while in one sense the whole of the Universe is extremely simple (multiples of the primary element, hydrogen), the simple becomes very complex when we probe into the constituent parts of the atom. We must give careful attention to the atom which is No.1 on the chart of the 103 natural elements. All of the matter of the Universe has been constructed out of it. "With the sole exception of pure hydrogen, all matter (including the matter from which we ourselves are made) must have originated in the center of stars through atomic nuclear fusion reactions" (von Ditfurth, *Children*, p. 11). He is referring to the creation of heavier-than-hydrogen elements which took place in the fiery nuclear furnaces which are found in the core of the stars in the Galaxies, using hydrogen as fuel. This kind of action is still going on.

We have mentioned before how small the atom is. The proton nucleus and the electron orbiter are, of course very much smaller. It is difficult to conceive of such small entities, if indeed they should be given such a name. Even more difficult to grasp, however, is the fact that scientists have been able to calculate the speed of the electron and have arrived at the conclusion that it travels around the proton 100 million billion times every second! This means that the electron moves around the proton so rapidly that for all practical purposes it surrounds it with a shell. This accounts for the stability of the atom. Whether hydrogen or iron, it makes no difference—it appears to be a "piece" of matter.

The "shell" concept, however, is very real for as the elements of matter are formed, the number of electron "shells" increases as the atoms become more complex. The diagram, adapted from *Basic Electricity*, by Abraham Marcus, illustrates the arrangement of the shells around the nucleus of the elements. The complexity and the order in the make-up of the atom is clearly demonstrated. The number of shells depends on the number of electrons in orbit about a specific nucleus of protons. The number of protons in the nucleus of an atom defines the nature of the element. One proton = hydrogen; two protons = helium; three protons = lithium; six protons = carbon; eight protons = oxygen; and so on up through Uranium, one of the heaviest elements, having 92 protons and 146 neutrons in its nucleus. Uranium is the element used to develop nuclear power. The Chart of Elements goes on to show a total of 103 elements, the latter ones being very unstable. For every proton in the nucleus there must be an electron in orbit. We will quote Marcus in his explanation of the diagram. "The shell nearest the nucleus (shell # 1) may contain up to two electrons. If there are more than two electrons, the excess form a second shell (shell # 2) around the first. This shell may hold up to eight electrons.

"After shell # 2 is completely filled, a third shell (shell # 3) is formed. This shell may hold up to eighteen electrons. However, the outermost shell of any atom may not hold more than eight

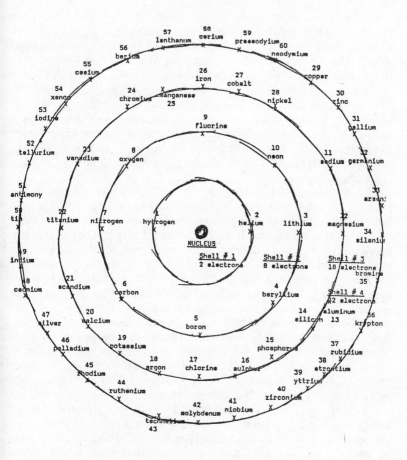

ELECTRON SHELLS AND ELEMENT FORMATION—The diagram indicates the Shell # and the number of electrons available to the Shell. The name of the element is given—with its proper number—indicating the number of protons in the nucleus of the given element together with the number of electrons in orbit about the protons of the nucleus.

electrons. Thus, after shell # 3 has eight electrons, shell # 4 is formed. Only after that shell has one or two electrons does shell #3 receive its remaining ten electrons. With shell #3 completely filled, additional electrons are added to shell #4. When the latter shell has eight electrons, electrons begin to appear in shell #5. The electrons are added to shell #4 until its quota of 32 is completed. In a somewhat similar manner, shells #6 and #7 are created. Only those electrons in the outermost shell of a particular atom (that is, those electrons with the highest energy levels) are involved in chemical and electrical phenomena. These are called *valence electrons.*"

The developmental order for the 103 natural elements (we speak of natural elements because, since scientists have learned how to split the atom, they have also learned how to create new, very unstable elements) is of fundamental importance to our understanding of the Universe. It destroys completely the concept that the Universe has developed by happenstance, without any Intelligence involved. By no stretch of imagination can the element-building system of "electronic shells" as has just been described, be attributed to chance, the random movements of non-matter energy particles, protons and electrons, either in the vastness of space or in the nuclear furnaces in the heart of stars. There is nothing random in the determination of the number of electrons in the shells. Neither is it happenstance that the outer shell can hold no more than eight electrons. Chance has no place in the holding action that occurs when a shell has reached the number of eight electrons and then does not accept more until an additional shell has been formed. It is not an accident that when the additional shell has been started, another holding action sets in until in the building process the unfilled shell is completed before continuing with the filling of the new shell. We should also point out that when an outer shell is complete, that is, it has all the electrons allowed on the outer shell, the resulting element is inert, it does not react with other elements to form new elements or compounds. Thus, the shell for the atom helium is complete when there are two electrons in orbit and helium is an inert gas. When the second shell is complete, the resulting element is the inert neon gas; the third complete shell produces the inert argon element. The chemically active elements are those with incomplete electron shells.

It is significant that elements very important to plant and animal life are found on the first two shells, almost clumped together. Hydrogen is on shell #1 and on shell #2 are found carbon, with four electrons on its outer shell, oxygen with six electrons and nitrogen with seven. These elements are involved in the hydrocarbon chain, and the amino acid group, essential for the life molecule. Two other significant elements for life development, sulphur and magnesium are located on shell #3. Should anyone seek to calculate the probability of the development of such a complex system of element building by accidental or random events, the lifetime of the Universe would prove too short!

As we have mentioned earlier, for every proton in the nucleus of an atom there must be an electron in orbit about the nucleus. "The atoms of one element differ from those of another element only in the number of their protons and electrons; it is this difference that makes an element—whether liquid, solid or gas, what it is," says Ralph E. Lapp in *Science Life Volume: Matter*. Lapp is the author of *Nuclear Radiation Physics* and a number of other volumes relating to the atom bomb and government policy in an H-Bomb era. What he says is that all matter in the Universe is basically the same stuff, just different configurations of protons and electrons. The subject becomes more interesting and complex when we realize that electrons and protons are both energy forms, particles or waves, one with a negative and the other with a positive electrical charge. Thus, they too, are basically the same substance—if we can use that word for an energy form.

The mystery of the atom deepens immeasurably when we are told by the scientists that the proton, tiny as it is, has been split into thirty-two distinct parts. "Atomic nuclei may release thirty-two "fundamental" particles when a high-energy beam of protons produced in a particle accelerator slams into a target." Half of these particles have a positive charge and half have a negative charge. These particles exist in free space for no more than a ten-billionth of a second and "Certain combinations of them, known as "Resonances" live for no longer than a hundred-thousandth of a billion billionth of a second!" (*Matter*, page 154).

Figures like these are, of course, meaningless to us. Nevertheless, they are real and if any of these tiny particles were not to be present and functioning, the whole chain or wave of Energy-

Matter would collapse and there would not be an atom or a Universe! *"Without them, the equations of nuclear events do not balance properly"* (*Matter*, p. 136, emphasis mine). Again, it is necessary to point out that the scientific evolutionist must be held to his turf, that everything has happened by sheer accident. Yet, when they bring our attention to the microscopic picture like that just presented, the concept of random development simply breaks down.

We have just seen that, theoretically, all matter of the Universe, the atoms of whatever element, can be reduced to a point of being only a form of energy, which is what Einstein said in his equation: $E = MC^2$ or, Energy = Mass times the speed of light squared. This also means that in some manner, Energy must be linked up with time, for that is what the "speed-of-light" is all about. Furthermore, as matter has been reduced to Energy in a relationship with time, it has become a non-material something, which can best be translated as "spiritual." This brings to mind the otherwise strange statement by the writer of the New Testament letter to the Hebrews, chapter 11.3. "Through faith we understand that *the worlds were formed by the word of God, so that things which are seen were made of things which do not appear.*"

We are forced to face the mysteries, the meaning of terms like radiation, pure energy, non-material, which have been used in our discussion in the realization that all the matter of the Universe is basically the same stuff, non-material radiation or pure energy turned into what we call "matter" through the ingenious organizational structure of the atoms of the specific elements. This brings us to write Einstein's equation as E = THE SPIRIT OF GOD + ACTIVE IN TIME. Time, as we know it, began with the birth of our Universe. Thus, as the Creator-God has exercised His will, He has done it in our universe time. But He, as the Eternal One—moved to do this in *His* time, which is apart from our time.

Instead of ruling the Creator-God out of the picture—pushing Him "off the edge of infinity," as Paul Davies suggests—we would provide a direct link between the Creator-God and the Mass of the Universe by placing Einstein's equation in a theological setting where E = The Sustained Expressed Will of the Creator-God = The Spirit of God + Active in Time. This concept could possibly resolve the problem between the theories of the Expanding Universe and the Static State Universe. It may well be that as the Creator-God continues to sustain His expressed will, He is thereby feeding into the Universe the energy to keep it going. Those who hold to the expanding universe theory should have no problem, for as long as the Creator-God sustains His will, the Universe will continue to expand as it is doing.

This concept could also resolve the problem the scientists have pondered with respect to the huge stars that go through what they call "gravitational collapse" and exit from the Universe by way of a Black Hole. (See chapter 7.) This exit of mass from the Universe upsets the law of the conservation of energy, for it allows energy to escape. If, however, the escaping energy is replaced by a constant flow of energy into the Universe by way of the sustained expressed will of the Creator-God, the problem is different but the equations should balance!

Thus, we suggest that the Creator-God in expressing His will brought into being the "worlds" as we know them—the Universe! The scientist in describing the origin of the Universe speaks of an enormous explosion of energy but he does not explain the source of that energy. He goes on to describe how the energy particles became "matter" as the electrons and protons formed a sustained relationship in the hydrogen atom. He follows this with the description of how all the Universe as we know it developed out of the hydrogen that was provided in the creation cloud. The description may well be correct but the origin of it all has not been explained. The scientist simply does not deal with the problem of the source of energy—the apparently inexhaustible energy of the electron is not explained.

It is our position, as we have indicated, that the "E" of Einstein's equation represents the continuing sustained expression of the will of the Creator-God, pouring out into the Universe as creating power. The substance of the Universe comes out of it. The electron, proton, neutron, photon and neutrino, are all energy forms flowing out of the energy of the Creator-God. Thus, the energy particles that appeared in the Creation-Cloud are not "things" nor are they truly particles. They are expressions of the continuing will of the Creator-God. To describe them as "energy waves" is possibly as close as we can come.

So, in the cloud of the Big Bang of creation, the protons, electrons, neutrons, photons and neutrinos were simply *projections of the sustained expressed will of the Creator-God.* He not only created them by the exercise of his will, but as long as he wills to do so, he sustains what he has expressly willed. He also put in place the Laws of Nature and under these Laws his Universe has been developed in all of its marvelous parts, including the Milky Way Galaxy, the Solar System and Life on Planet Earth! This, as we stated earlier, is consistent with all that the scientists have discovered.

The scientist asks us to accept on faith his hypothesis about how the Universe came into being. "The first theory places the question of the origin of life beyond the reach of scientific inquiry. It is a statement of faith in the power of a Supreme Being not subject to the laws of science. The second theory is also an act of faith. The act of faith consists in assuming that the scientific view of the origin of life is correct, *without having concrete evidence to support that belief" (Jastrow, Sun,* pp. 62–63). The entire proposal of the scientific evolutionist is built on the acceptance of the second theory. It represents a leap into darkness! Furthermore, as we have seen, the design of the atoms and the structure of the elements indicates both design and purpose. The absolute order found in the Universe which has enabled the mind of man to discover the basic laws of physics which govern the Universe, delivers a death blow to the hypothesis of the scientific evolutionist. It is strange that the scientist should ask that his hypothesis should be accepted on faith while refusing himself to accept on faith the proposal that a divine Intelligence, the Creator-God, is responsible for all that the scientist has discovered. No scientist that I have read states this more strongly than Paul Davies, in *The Edge of Infinity*: "It is truly remarkable that the religious adherents have not learned the lesson of history that nature can order its own affairs. The great mistake of theology is to cling to the prehistoric belief in an organizational-manipulative diety. All along they seem to have missed the point that the true splendor of the cosmos is not the initiation of the organization, but in the laws of nature that nurture and sustain that organization and operate the cosmic system in an ordered way. What is the virtue of a god, beyond the edge of infinity, who is not a part of the stunning beauty contained in nature's mathematical laws?" (p. 171).

What Davies is really saying here is that we should ignore the failure of the scientist to come up with an answer to the problem of the origin of the Universe—that we should accept his position that the "initiation of the organization"—that is, the creation of the Universe—is not important but that rather, "the laws of nature that nurture and sustain that organization" are the significant factors. In this, he simply ignores the fact that he has not explained the existence of the "laws of nature" that are evidently the necessary ingredients for the functioning of his universe. He ignores the fact that an "energy source" must be identified or "laws of nature" will not function, He ignores the fact that he leaves unexplained the "initiation of the organization"—the creation of the Universe.

Thus, our picture has become more clear. The Creator-God willed that there be a material Universe, such as we have. The expression of His will moved with the speed of thought. He said, "Let there be"—and there was! What is thought? It is an expression of the "mind" and as such has a kind of "being." The thought of the Creator-God as expressed produces energy in the Universe and explains the continuous energy of the electrons and the other fundamental energy waves. Thoughts are like pellets of Energy-Matter being hurled by the mind (the will) of the Creator-God into the continuing Universe. The thought of all humans is in the same manner a form of Energy-Matter, as the parapsychologist would agree. Does this mean that the Universe will continue forever? not likely, for the purpose of the Creator-God will at some time be fulfilled. Until that time, however, the Universe will continue to exist.

By His sustained expressed will, He called into being the components of what our scientists have called the "fireball." These components were, to use the scientists' own term, "non-material"—pure energy forms—which in accord with the Laws of Nature, as established by the Creator-God formed the atoms and in so doing brought into being real matter as we know it. This "calling into being" of the energy components was done in eternal time, before the universe time began. When the Creator-God said the word, the Fireball exploded the Universe into being and what has happened since is history. The scientists have described it well.

CHAPTER V
THE FORMATION OF THE GALACTIC SYSTEM

We have come to the end of the beginning! The Creation-Cloud of the Big Bang has cooled to a point when electrons are able to go into sustained orbit around a proton and the atom hydrogen has been formed. The entire Creation-Cloud was at this time changed into a huge cloud of hydrogen gas—75% and helium gas 25%. There may have been traces of lithium and berylium. These would be the only elements of matter in the Universe until after the development of the Galactic System in which stars would become the factories which produced new elements. As we shall see as our account goes on, there are problems with the view of the scientific community with respect to the creation of the elements. In view of the very disciplined manner in which the atoms of the elements are made up, it is difficult to accept the scientific view that their production occurred in a happenstance manner in the random explosions of huge stars.

The diagrams which follow illustrate in a simple manner the make-up of atoms of matter, showing that what we think of as solid stuff is really a carefully designed arrangement of non-material energy forms. We will follow through the process of element development as described by Marcus in chapter Four—that is, through the first two electron shells.

The Make-Up of Matter

The sources are non-material energy forms for which we will use symbols.

○ = an electron, an energy form found in the original Creation Cloud; carries a negative electrical charge; it has almost no mass or weight, but we will give a 1 ugf or unit of gravitational force. This will help us to understand the structure of atoms of elements. At a later point we will explain the term which we have invented.

● = a proton, also an energy form found in the Creation Cloud; it carries a positive electrical charge; it has a mass or weight 1836 times that of an electron, so we give 1836 ugf.

⊗ = a neutron, also an energy form found in the Creation Cloud; it carries no electrical charge and is believed to be a joined electron and proton; it has a mass or weight a very tiny bit more than a proton or 1836 + 1 for the joined electron, or 1837 ugf.

These are the energy forms that are the building blocks of all matter; all the elements are made up of different configurations of these forms.

⊙ = *hydrogen*, composed of one electron orbiting 1 proton. It is the first material element. It has no electrical charge because the negative electron and the positive proton cancel out. All atoms of elements are thus neutral as the atom of every element must have one electron in orbit about the nucleus for every proton in the nucleus. There is a possibility that under certain conditions an atom may be stripped of an electron from the outer shell, or it may take on an extra electron which then gives the atom an electrical charge and the atom is said to be "ionized." The weight or mass of the hydrogen atom must be 1 + 1836, or 1837 ugf. Refer to chapter 8, page 129 and to chapter 10, page 151, for the explanation of the term "ugf"—for it is our term, not taken from scientific texts.

The electron orbits the proton at a rate of 100 million billion times per second, we are told in *Matter*, page 125. The speed is so great that it appears as if the electron has surrounded the proton in a shell. We read in *Matter*, "It is this everywhere-at-once character of the electron in orbit that makes for the solidity and rigidity of the atom's structure."

If 1 neutron were added to the nucleus of a hydrogen atom we would have an atom of "heavy hydrogen" with 1837 + 1837 = 3674 ugf. The neutron adds mass to the hydrogen atom but does not change the nature of the element. The heavy hydrogen is called "deuterium". If two neutrons entered the nucleus, we would have "tritium" = 5511 ugf.

If we were to picture an atom of hydrogen we should imagine a proton as large as a golf ball. The electron in orbit around it would then have to be one mile out! Thus, it is easy for a neutron, the size of a proton, to slip into the nucleus of an atom of hydrogen—or any atom. The number of neutrons in the nucleus of an atom plays an important role in both element formation and in the development of matter where compounds are assembled. They are significant also in both the fusion and fission processes where elements are built up and broken down.

Helium, the second material element, composed of two electrons orbiting a nucleus of two protons. As the helium atom functions in its element-formation role, it has two neutrons in its nucleus—so we diagram that form also. The electron shell for hydrogen and helium allows for only two electrons. Thus, with the two electrons, the shell is "full" and helium is an inert element. As has been said, all matter is made of the same stuff—the electron and proton non-material energy forms. Material substance is thus made from non-material energy forms that enter into a sustained relationship.

Lithium, the third material element, composed of three electrons orbiting three protons. As the first shell can have no more than two electrons, a second shell develops as the third electron goes into orbit. This shell has room for eight electrons according to the design for atomic elements as shown in chapter Four. Thus lithium, with only one electron on its outer shell, is a chemically active element. It will easily give up its one electron to form a union with another element to form a third element or it will take on electrons to form new elements. It will also combine with other elements or compounds to form other chemical compounds.

Beryllium, the fourth, composed of four electrons orbiting four protons. In the first shell there are two, leaving six empty electron spots. Thus beryllium is a chemically active element.

Boron, the fifth, composed of five electrons orbiting five protons. On the outer shell there are three electrons, leaving five empty spots. Boron is thus a chemically active element.

Carbon, the sixth, composed of six electrons orbiting six protons. On the outer shell there are four electrons, leaving four empty spots. Carbon is a chemically active element that has an equal capacity to go either way in reactions—either to take on four electrons or to share four electrons. It is thus an active element. Carbon is the key element in the hydro-carbon chain which is so significant in the development of organic matter.

Nitrogen, the seventh composed of seven electrons orbiting seven protons. On the outer shell are five electrons, leaving three empty spots for the making of amino acids which play a significant role in the development of the "life molecule". 78% of the atmosphere of the Planet Earth is made up of nitrogen. Nitrogen is thus a chemically active element. Nitrogen is necessary for the making of amino acids, etc.

Oxygen, the eighth, composed of eight electrons orbiting eight protons. On the outer shell are six electrons, leaving two empty spots. Oxygen is thus a chemically active element. Oxygen is absolutely necessary for both plant and animal life, as carbon dioxide for plants and free oxygen for animals.

Flourine, the ninth, composed of nine electrons orbiting nine protons. On the outer shell are seven electrons, leaving only one spot empty. Thus flourine is a chemically active element.

Neon, the tenth, composed of ten electrons orbiting ten protons. There are eight electrons on the outer shell, leaving no empty spots. Neon is therefore an inert element.

Sodium, the eleventh, composed of eleven protons in its nucleus, with two electrons on shell #3. And so it goes, all the way up to the heaviest of all the natural elements, having 103 protons in its nucleus. The uranium isotope used in the nuclear processes is U238—which means that there are 92 protons and 146 neutrons in the nucleus. We will not carry the illustration farther. It is evident that the elements are all made up of the same electron-proton energy forms and that the *system* is one of order and design. We do not imply that the 103 elements were created in sequence, but we do insist that they must all be accounted for and that their existence is not the result of happenstance accidents.

The electrons, in addition to orbiting the nucleus, have a spin. The spin of the electrons on the same shell must always be in opposite directions, (*The Nature of Matter*, Otto R. Frisch, page 115). We mention this to point out that design is built into every part of the system that translates energy waves into the matter of elements. There are no accidental happenings involved in the creation of matter. The Laws of Nature are in place and operating. The Creator-God willed the Universe into being. He did so in His eternal time and set up the Laws of Nature to be operative with the beginning of universe time. This may well be what St. Augustine meant when he said, "The world was made with time and not in time."

Davies makes a wrong assumption when he suggests that there would be little value in a god who, after creating the Universe would then step out of the picture and let "the system run itself according to the laws of nature." When we read the first few verses of the Biblical account we get a much different picture. "In the beginning, God created the heaven and earth. And the earth was without form and void, and darkness was upon the face of the deep. And the Spirit of God moved upon the face of the waters. And God said, let there be light and there was light." The picture we receive from these words portrays a world that was dark, empty of all life, chaotic in the turbulance of the creative period. The scientists' description of how the planets of our Solar System were developed fits the picture. But, the Creator-God was in charge then and always has been. There is no evidence in the Bible that the Creator-God abandoned His creation at any time. We will discuss this subject more completely when we come to the section of this study that will deal with the eternal, transcendental being of God, who with His angels and archangels worked in eternal time before universe time and continues to work now in universe time within eternal time.

Again, we must remind ourselves that the scientific evolutionist has chosen the turf on which to stand as he defends his thesis that everything has come into being by happenstance. Davies joins the group with strong statements in *The Edge of Infinity*, where he pushes the Creator-God "off the very edge of space-time." Having so dealt with Him, he states: "The more believable image [of the creation of the Universe] is that of the chaotic singularity, where the breakdown of law leads to complete randomness, so that the emerging material and influences have no in-built organization at all. Any emerging structure is purely accidental and exceedingly improbable" (p. 169).

Davies then tries to prove his case by asking for evidence that the "big bang singularity [which he here is equating with the Creator-God] was responsible for creating the high degree of order that we observe today in the Universe." The obvious is evidently entirely unobvious! The very point that he makes with respect to the "high degree of order that we observe in the Universe today" speaks to the point of an Intelligence, a Creator-God, who brought order out of the previous chaos. He points out quite accurately that in the beginning "there were no galaxies, no stars, or planets, no people, no atoms, and not even atomic nuclei". . . . There is no apparent need for a supernatural organizer—the laws of nature themselves seem capable of generating the present high order of structure and organization that makes the universe so interesting" (p. 170). He could not have stated it better! This is the key that the scientist has overlooked. The Laws of Nature were set in place before the Big Bang by the Creator-God, and were there, ready to receive the particles-waves of non-material radiation which were certainly in a chaotic state and gradually over the first general period of universe time, bring them into the order called for. Davies has simply misread the beginning verses of Genesis. He says, "The old idea of a sort of 'package universe' set up in cosmic splendor, does not accord with the evidence." Davies does not appear to understand that the Bible does not support a "package universe" but rather tells of a creation that in the beginning was without form, void and dark. The form, order, together with light, came later as the Laws of Nature exercised their powers under the guiding influence of the Creator-God and His unnumbered hosts of workmen.

At the risk of coming down too hard on one man, but a man who is a sense speaks for the scientific community, we quote Davies again: "This conclusion completes the steady retreat suffered by the concept of an organizational-manipulative deity that began *two or three millenia ago.*" He has been trapped by the errors of Bishop Ussher! He continues, saying, "The organizational-manipulative God has been displaced right back through time, *even off the edge of time*, banished to a domain beyond the natural world . . . The cosmos as now ordered has arisen automatically from primeval chaos."

The position of the scientific evolutionist cannot be put more clearly. The Universe as we know it, they say, has come into being without the aid of any supernatural being or Creator-God. It simply happened by pure accident and the order we see in the Universe came into place automatically in accord with the laws of nature. No explanation is given for the existence of the lower case laws of nature! No explanation is given for the coming-into-being of the energy particles-waves in the Big Bang of creation. No explanation is given for the design and order found in the development of the atoms of the 103 natural elements out of pure energy. No explanation is given as to how pure energy, as a nonmaterial substance, became a solid, liquid or gaseous material substance. The scientists simply gloss over these matters as though they were of no concern. This is not in accord with the precepts of the scientific method. Happenstance developments occuring randomly over billions of years, do not result in the kind of order and design which we find in the Universe.

Since we have come to the end of the beginning, we must move on to what happened after the beginning, when the huge amorphous creation-cloud of radiation had expanded and cooled to the point when the first signs of order in the Universe became evident. The cloud, now composed of hydrogen and helium, began to break up. Apart from the fact that the Universe had a beginning, this is one of the early pieces of evidence that the Universe is not infinite. If it were infinite, there would have been no space available for the billions of the new clouds to move into. The idea of Gribbin, that it was the expansion of Space that "pulled the creation-cloud" out with it is open to question. The space was there in advance and the original cloud broke into smaller clouds and moved into it. There was not only "some" space but so much that these clouds, continuing their outward movement, were soon separated from each other by vast distances. It has been calculated that at present they are on the average twenty million trillion miles from one another—and this in every direction! We now call these clouds the Galaxies. Each galaxy has developed within itself and out of the substance it brought from the creation-cloud, from 50 to 200 billions of stars. The Galaxies, as they were originally composed, were made up of the hydrogen and helium gases of the creation-cloud. As time passed, billions of years, they also held clouds of dust (galactic debris) likely some of the elements which had been created in the cores of early stars that exploded. But, we are getting ahead of our story.

The break-up of the original creation-cloud began roughly at about one billion years after the Big Bang, according to Jastrow. It happened over a long period of time. In a sense, it is still happening in that the scientists state that the galaxies continue to move away from each other at an ever increasing speed, the ones farther out traveling at a possible speed of a billion miles per hour. The farther away a galaxy is from our Milky Way, for instance, the faster it is supposed to be moving from our Galaxy. There are problems with this theory which we will address later in this chapter.

Weinberg, in *The First Three Minutes*, states: "The differentiation of matter into galaxies and stars could not have begun until the time when the cosmic temperature became low enough for the electrons to be captured into atoms.... This is not to say that we actually understand how galaxies are formed. The theory of the formation of galaxies is one of the great outstanding problems of astrophysics, a problem that today seems far from solution" (p. 74). Maffei, in *Beyond the Moon*, agrees with Weinberg that until the temperature had fallen drastically, the formation of stable matter was not possible. "It is at this point (about one million years after the Big Bang) that, from the primordial mixture of hydrogen and helium, the first condensations begin to form, from which will later be born the galaxies, populated by stars, in whose interiors (as we have seen) are formed from hydrogen, the heavy elements, up to iron, lead, gold" (p. 323).

We will see, as our study continues, that the problem of the existence of all 103 natural elements has not been addressed adequately by the scientific community. The explosion of the super nova stars, which we will discuss later, is supposed to have thrown out into space the elements they had formed in their cores. But this is an iffy answer to the question. How many of the elements were created in this manner? Given the great distance between the stars, how could one star-cloud pick up the elements "floating" in the vastness of space? And, getting down to specifics, how could one star-cloud, our Solar System, manage to pick up all 103 of the elements—and in quantities large enough to be meaningful and then deposit them on one planet we call Earth?

As long as the scientist clings to his position that everything has happened by chance, by the random movements of whatever was in the original creation-cloud, he will find the answers to his problem difficult to discover. On the other hand, for those who believe that the evidence points to the sustained expressed will of the Creator-God, the answers can be found. In the design for the development of the Universe, it was necessary that the galaxies be formed and that within the galaxies, the stars should be developed. No new elements beyond the first two would be created until the nuclear fusion process within the cores of the stars was in operation. A significant fact is that while the galaxies are continually moving "outward" at great speed and are thus expanding the Universe, the galaxies themselves are not expanding. When they separated out and the original cloud was dispersed, they stopped their individual expansion while they continued their outward thrust. Dr. Richard Morris, astrophysicist and executive director of COSMEP, writes in his book *The End of the World*: "Only the galaxies move apart (in the expanding Universe) not the stars within them. Neighboring stars remain approximately the same distance from one another; each galaxy is held together by gravitational attraction" (p. 129).

The formation of stars began within each galaxy cloud as it left the creation-cloud. This tells us that each galaxy cloud formed a core, a gravitational center, and thus controlled the movements of the newly forming stars . . . much like we will see happened in the Solar System Cloud when our Sun was formed from our Solar System Cloud as it separated out from an arm of the Milky Way Galaxy Cloud. Our Sun, the core of the Solar System Cloud, controlled the movement of the forming Planets. The core of the

Galaxy controlled the movements of the Suns or Stars of the Galaxy. Carrying the comparison of the development of the Solar System with the development of the Milky Way Galaxy a bit farther, we would say that the galaxy clouds that separated out from the Creation Cloud to form the galactic system very likely contracted as they formed gravitational centers. At the center of each galaxy there appears to be a gigantic nuclear furnace, similar to but very much larger than that of any star. This means that each galactic cloud likely went through a contraction-heating-cooling-expansion process similar to that of the stars during their formation period. This we will discuss in the following chapter. This process was an absolute necessity, to achieve the temperature required to trigger nuclear reactions powerful enough to generate the raw power needed to hold a galaxy together!

Maffei has taken his fellow travelers with him on an imaginary journey to the center of the Milky Way Galaxy. It could have been any galaxy, of course. They have determined that the center lies in a region that has a diameter of about fifteen light years, or about ninety trillion miles. "At the center of the region of 15 light years in diameter, there is an intense source 100,000 times brighter than the sun. It appears to have a diameter no greater than 170 times that of our solar system and a temperature of 2,000 degrees K. This, then, is what lies at the center of the Galaxy: an enormous furnace, a huge crucible . . . We have discovered an immense furnace but a new mystery as well. For we do not know how that furnace came to be there, nor what it is burning, nor how. If we only knew, we might perhaps have solved not only the mystery of the structure of the Galaxy, but also the mystery of its origin and evolution, and perhaps even the origin and evolution of the Universe" (pp. 235–236).

While there certainly are unanswered questions, some are answered. The center of the Galaxy has the same characteristics as the center of a star, except on an entirely different scale. It is a nuclear furnace. Also, as the Galaxy is known to be rotating about the center of its gravitational field, this furnace is that center. As the core of our star supplies the gravitational energy for holding the Solar System in orbit, it could appear that this galactic center holds all of the stars of the Galaxy with their solar systems in orbit around itself. It thus is the power-house of the Galaxy. Finally, to this observer, it points to the logical conclusion that the Galaxies, all with similar nuclear furnace centers, are also in orbit around their gravitational center which, as we will indicate later in this chapter, must be located at the point where the Big Bang exploded, the center of the Universe.

"For us, the important point is that in the early universe, at temperature above 3,000°K, the universe consisted not of the galaxies and stars we see in the sky today, but only of an ionized and undifferentiated soup of matter and radiation" (Weinberg, p. 75). It is significant that the light particles-waves were in a sense, bound during this period. The photons were not able to penetrate the ionized gas and thus even though the Creation-Cloud contained enormous numbers of photons, all was darkness for the photons could provide no light. They were neutralized by the electromagnetic field set up by the negatively charged electrons and the positively charged protons. When, however, the atoms formed, the electrical charges were neutralized and the electromagnetic field vanished. The photons then for the first time were able to flood the Universe with light. "At the same time (when the first atoms formed) the obscuring fog of radiation cleared up, and the Universe suddenly became transparent. The reason for this change was that light, which is a form of radiation, cannot pass through electrically charged particles, such as electrons and protons. However, atoms which are electrically neutral, do not block radiation appreciably. As soon as the electrons in the Universe had combined with protons or other nuclei to form atoms, rays of light were able to travel great distances unhindered, and it became possible to see from one end of the Universe to the other" Jastrow, *Astronomers*, p. 128). No one was there, of course, to see the Universe, but the photons were unleashed to illuminate it!

With the formation of the Galactic Clouds, the process toward the development of an orderly Universe picked up speed. The movement of the massive clouds was rapid for they were still under the influence of the "outward thrust" of the original explosion. In addition to that thrust, the force of gravity became weaker as they moved out into space—away from the "center" of gravity. The scientific community is quite ambivalent about such a "center" for the Universe but it appears to this writer that if such a center were not to exist, the whole concept of gravitational force would have to be scrapped.

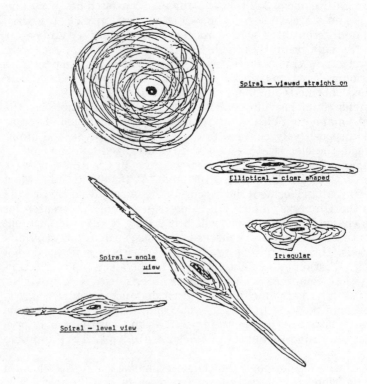

SPIRAL, ELLIPTICAL AND IRREGULAR GALAXIES

The Galactic Clouds were composed of hydrogen and helium—the only elements that existed at this time—except for trace amounts of lithium and beryllium and the non-material energy particles-waves of neutrons, photons and neutrinos. While these clouds may have appeared to have been chaotic and without order, they were nevertheless acting under the Laws of Nature. "The depths of space are ruled by the same laws and forces which govern terrestrial life" (Ditfurth, *Children*, p. 12). In *The Nature of Time*, Whitrow states that there is "Powerful evidence that the system of galaxies forms the framework of the Universe" (p. 151).

As has been indicated, the actual formation of the clouds which separated out from the original Creation-Cloud is not really understood by the scientific community. We will let the matter rest, knowing that the system did form, that it exists and that in one of the Galaxies, our Sun was born and our Solar System was developed. All galaxies, however, are not the same in structure and function. This we will touch on now, for in this differentiation between the galaxies we see another kind of design or order in the Universe. We may not yet know the significance of the design or the purpose, but we must recognize that it exists. There are three basic classifications for the Galaxies in the Universe, the Spiral, the Elliptical and the Irregular. They all are very large, averaging about 600,000 trillion miles in diameter. On average, each contains from 50 to 100 billion starsplus huge clouds of hydrogen gas (except for the elliptical), other gases, debris from exploding supernova stars and other substances. New stars are being continuously created in the spiral galaxies which have the hydrogen clouds. The elliptical galaxies do not have hydrogen clouds and do not appear to give birth to stars. The stars in the spiral galaxies represent both old and new stars which are classified as Population I and II stars. More will be said of these when we discuss the life of a star. The irregular galaxies vary greatly in shape. The spirals appear to be discuss-shaped while the elliptical appear cigar shaped or like the spirals without the bulge. The diagrams to the left give a picture of about how they look. Note the X in the spiral galaxy. It represents about the location of the star which is the Sun of our Solar System. The irregulars may or may not give birth to new stars. Actually, the Galaxies are so far apart that it has not been possible to study them in any detail. The Galaxy closest to our Milky Way is the Andromeda—12 million trillion miles away, or 2 million light years! Until a century ago, it was believed that our Galaxy constituted the whole of the Universe. Now, with the aid of new instruments and techniques, and a greatly advanced knowledge in the field of astrophysics, the astronomers have sighted galaxies as far as 12 billion light years away or more than 72 billion trillion miles. It is believed that the edge of the Universe will be found to be about 15 billion light years out from our place of observation, or 90 billion trillion miles. We cannot grasp figures so large, but they are real.

"Astronomers believe that the total number of galaxies in the universe far exceeds the number of suns composing each of them (from 50 to 200 billion)" (Ditfurth, *Children*, p. 40). He also said on page 26, "A galaxy is an independent stellar system, lens or discuss shaped, which rotates around a single center of gravity." "The cosmic swirl of stars composing a galaxy is inconceivably vast, so vast that even the oldest of them have rotated only about twenty times around their own axes in all of the 12 to 15 billion years since the Universe began! And yet, these galaxies have been rotating at such speed that their swiftest moving suns cover more than 300 miles per second. . . . Galaxies are the largest formations in the Universe" (p. 31). "The Galaxies began to form when the Universe was roughly one billion years old. The formation of stars probably began shortly after the formation of the first galaxies". (Jastrow, *Astronomers*, p. 129). "Our Galaxy is but one insignificant stellar island among thousands of millions scattered across the sea of space" (Gribbon, p. 17).

Thus, while we do not know much about the Galaxies, we do know that the galactic system began to develop about one billion years after the Big Bang of Creation. The split-up of the original Creation-Cloud did not happen instantly but took place over a great period of time. Thus, the astronomers can speak of some galaxies as being "older" than others. We know, too, that each galaxy is an independent stellar system that rotates around a "single center of gravity." This means that they turn on their own axes—like the Earth does. The manner in which the Galaxies are in motion in space is not really made clear by the astronomers. They do move and the movement is supposed to be "outward"—that is, in an expanding Universe, they are moving away from each other. Nothing that I have read would indicate that the Galaxies are in orbit about a gravitational center. As I have indicated earlier, this would mean that they break what appears to be a universal law, that each body in space rotates on its axis and orbits the center of its gravitational system. I would believe that the Galaxies do the same. They could also be moving away from each other by the enlargement of their orbit due to the outward thrust from the Big Bang and to the decreasing effect of the pull of their gravitational center. This concept, however, runs into conflict with the views of the scientific community, which does not admit to a "center" for the Universe. A number of illustrations have been developed to explain the concept that the Universe does not have a center, but they are not convincing. One uses a loaf of raisin bread. When a loaf, rises, due to the action of the yeast, the various raisins are pulled along with the enlarging mass of dough. Thus each raisin is said to be "moving away" from each other raisin and the outermost ones move faster and thus farther than the inner raisins and no raisin can be said to be at the center. This, the scientists say, is a description of the expanding Universe, the Galaxies being the "raisins"! To me, this is an exercise in sophistry. No matter how large that loaf becomes and how far the outer raisins are from the inner raisins, there is still a "center" raisin and it has not really moved relative to the others.

Why the big fuss over this matter? It appears to an outside observer that the scientist, having lost the battle in the matter of the "beginning creation" of the Universe, does not want to admit to a "Center," which could imply the location of the Power that runs the Universe! In other words, to accept the concept that the Galaxies are in orbit around a common center of gravity would destroy their position that there is no "Source-of Power" at the center of the Universe—no Creator-God, or as Davies has put it, no "organizational-manipulative diety."

Another troubling thought must also be faced. The Andromeda Galaxy, our closest galactic neighbor, only 2 million light years away (12 million trillion miles!) and the only galaxy that our scientists have been able to study, *is not moving away from the Milky Way Galaxy*. In fact, the two are on a collision course! "Of all the spiral galaxies, Andromeda is the nearest to the Milky Way. The light from it is shifting toward the blue, not the red, showing that it is approaching us. The Andromeda galaxy is an exception to the rule that the galaxies seem to be receding; it and the Milky Way are gravitationally bound" (Morris, p. 108).

Does this mean that our Galaxy is going to be destroyed by a collision with the Andromeda Galaxy? Not likely, in Morris's opinion. "Galaxies can and do collide with one another. Astronomers have observed such collisions taking place. However, such collisions are not especially catastrophic events. If the Milky Way were to intersect another galaxy like it, the two would simply pass through one another. Again, the wide separation of the stars would make destructive occurrences improbable" (p. 86). Morris goes on to state that, should the Andromeda galaxy be made up of anti-matter, the story would be different—they would annihilate each other. "If such an event was to take place, however, it would lie billions of years in the future, long after the earth had been vaporized by the sun" (p. 88).

However, if the only galaxy near enough to be observed is moving toward the Milky Way, how can astronomers be sure that their theory that the galaxies are all moving away from each other is correct? The Andromeda and the Milky Way galaxies are only two among the hundreds of billions in the Universe. And these two are not behaving in accordance with the accepted theory relative to the expanding Universe. Does this not raise questions with respect to the movement of the Galaxies which the "Raisin-Loaf" analogy cannot explain? Also, if Galaxies are indeed colliding (passing through) one another as Morris stated, there are obvious problems with the theory that they are all moving away from each other.

Having accepted as fact the concept that the Universe as we know it started at the moment of the Big Bang, it appears that scientists must also agree that there must have been a moment either in time or out of time, and a point either in space or out of space, when and where this event took place. That moment would be the beginning of time as we know it and the beginning of the Universe. That place would be the gravitational center for the Galaxies of the Universe, and also the point where the Creator-God was exercised in the explosion we call the Big Bang. It may well be that the place is out of the time-space fabric of the Universe, but it does exist. This concept in no way limits the

Creator-God to that place for He is everywhere present and Universe Time as we know it does not apply to *Him*.

Returning to the Galaxies and their classifications, we go to Gribbin's description. Speaking of the classifications of the three basic galaxy types, referring to the original descriptions made by Edwin Powell Hubble, who is listed as one of the Great Ones among the astronomers, "His classification remains today, somewhat modified, and at the heart of it is the distinction between the elliptical galaxies, which do indeed look like ellipses in astrophotographs, and spiral galaxies, which are made up of a central bulge of stars rather like a small elliptical galaxy and a surrounding disc of stars and other material through which spiral arms can be traced. Galaxies which fit neither category are called irregulars."

The purpose of the elliptical and irregular galaxies has not become clear. They are large, containing the usual number off stars—from 50 to 200 billion—and they rotate on their axes and move in the "Great Dance" of all the galaxies in space. They do not appear to produce new stars—at least it is quite certain that this does not happen in the ellipticals. The latter have a greater number of very old stars—on the way to a cold death when their nuclear fuel is spent, or to a fiery explosion if they are of the size that demands this.

The spirals are just what the name implies, a formation of stars in which the stars-in-formation are located in what looks like spiral shaped arms swirling in the space occupied by the galaxy. The diagram illustrates this. In between the arms are huge spaces where fully developed stars, like our Sun with its Solar System are located. Also, in the spaces between the arms there are great clouds of hydrogen and other gases, dust (possibly containing much carbon) and what is called debris (possibly various elements which come from exploding supernova stars). The spiral galaxies are also called discus-shaped, as the diagram suggests. "The important point which does have to be explained by the new model (the concept that has developed since the latee 1970's) is that 75% of observed galaxies are spirals (including our own Galaxy), 20% are ellipticals and 5% are irregulars" (Gribbin, p. 41). Gribbin further indicates that while this is true for the observed galaxies, the numbers could be about equal if all the Universe could be observed. On the other hand, if the ellipticals do not produce new stars and if they now show a very large number of "old" stars on the way to extinction, would it not be reasonable to believe that the true ratio is about as shown among the observed galaxies? What significance this may have, I do not know—except that it would suggest that the mission of the elliptical galaxies may be about over or that they may be in a state off "change-in-function."

Our concern is with the spiral galaxy for it is here that we in our Solar System are located—in the Milky Way Galaxy. In the diagram of the galaxy types we have placed an X in the area where our star is likely located—about two thirds off the way out from the center of the galaxy toward the outer edge. It has been learned that there are two types of stars in the spiral galaxy we call the Milky Way. They are called Population I and II stars. The young, new stars, possibly only 100 million years old and still located in the arms of the galaxy, are the Population I stars. The Population II group consists of the older stars, such as our Sun, which is about four and a half billion years old. Old as it is, our Star is looked upon as being relatively young as compared to the age of the galaxy and the Universe. These stars have moved out from the arms of the galaxy into the space between the arms. It is important that we keep this in mind when in the next chapter we deal with the birth of our Solar System and our Planet Earth.

The second concern we have in discussing the classification of the Galaxies is to point out again that we have evidence of design in the development of the Universe. If there were no such design and if everything had come into place by happenstance, we would not find the "systems" of cosmic bodies that would be subject to classification. Also, we would not find that, as Ditfurth states in *Children of the Universe*, (p. 12), "The laws and forces ruling this Universe justify our calling it a 'Cosmos'; a harmonious entity where every part is responsive to every other." It is true that we do not have all the answers to the problems but it is also true that the design of the Universe cannot be denied. Design indicates purpose, and the two together indicate goals and an Intelligence, a Creator-God who is in charge.

It is important for us to realize what we are discussing at this point. We are talking about the ground we walk on, the food we eat, the bodies that enclose our spirits. The Planet Earth, with its rivers and oceans, its mountains and valleys, is a tiny part of the matter we are concerned with. When you stand on your high point and look up at the stars, you can see only about 6,000 under the best of conditions—but there are billions and billions more that cannot be seen. These too, are a part of the whole of the Universe! When the Big Bang happened, the whole of the Universe came into being—not as we see it now, composed of earth, stone, forests, moons, suns or galaxies, but as energy, intensely hot radiation, a kind of boiling plasma moving out into space.

For the first million years, or so, there was no matter or atoms of any kind for it was to hot for stable atoms to exist. Then, as the mass cooled, mainly due to the rapid expansion into a seemingly unending volume off space, the first atoms of hydrogen and helium formed. There were also the photons or light particles-waves and the mysterious neutrinos about which so little is known. And that was all—unless even these energy forms were broken down into what we call "fundamental particles" which cannot be subdivided. It was not until the original Creation-Cloud broke up into what we call the galactic clouds that in time, the stars were formed. And these stars were composed of only hydrogen and helium—gas! It is difficult for us to think of our Sun as being simply a mass of gas—but that is what it is. According to the theories of the scientific evolutionists, the stars became the factories which, in nuclear furnaces located in their cores, produced the other elements of the Universe. Maffei, in *Beyond the Moon*, writes; "In the first few minutes, through a marked synthesizing activity of the atomic nuclei, the light elements are born and a mixture is formed, composed essentially of hydrogen (around 75%) and helium (about 25%) and containing traces of other elements such as lithium and beryllium. Meanwhile, the temperature continues to decrease; after ten thousand years it has reached 10,000 degrees Kelvin, and after a million years it is barely 600 degrees. It is at this point that, from the primordial mixture of hydrogen and helium, the first condensations begin to form, from which will later be born the galaxies, populated by stars in whose interiors (as we have seen) are formed, from hydrogen, the heavy elements up to iron, lead, gold" (pp. 322f).

The elements that were formed were kept locked up in the interior of the stars until the time came when certain stars—only the larger ones—exploded and shot out into space some of the elements they had produced. This material then got caught up in the swirling clouds of hydrogen gas and in time, a new population of stars containing some of the new elements were formed by the same process that had formed the original stars. But this should be be told in the account of the life of the stars in the following chapter.

How many star explosions were necessary before one star-cloud four-and-a-half billion years ago was able to gather into its system all of the 103 natural elements is an interesting question. We will deal with it in the chapters that follow. We may then meet up with difficult problems. One thing appears certain, however; these 103 elements did not come together by accident!

CHAPTER VI
THE BIRTH OF THE SOLAR SYSTEM

"Nearly five billion years ago, in one of the spiral arms of the Milky Way Galaxy, a cloud of gaseous matter *formed by accident* out of the swirling tendrils of the primal mist. The history of the solar system began in that softly glowing nebulosity, bathed in the black light of nearby stars" (Jastrow, *Until the Sun Dies*, p. 54). Jastrow emphasizes that it happened by accident, no design or purpose were involved. Our Sun, with her nine planets, with their moons and satellites, were to come into being as this giant cloud within a much larger galactic cloud began to contract as it moved on its way to becoming the Solar System.

To develop such a system, this cloud would have to bring into being a star which we would call a Sun. This sun would have to have a mass great enough to control the planets and all else that would be formed within the Solar System. In *God and the Astronomers*, Jastrow states, "In the *random motions* of such clouds *atoms somehow come together by accident* to form small, condensed pockets of gas. *Stars are born in these accidents*" (p. 105). The concept runs like this: in the swirling gas of the cloud, two hydrogen atoms collide and stick together. This forms a polyatomic structure twice as heavy as a single atom. The law of gravity takes over and the "heavier clump" of hydrogen attracts another atom, and another. As the size of the structure—which will become the core of the cloud—becomes larger, so also its mass increases. This means that it will attract more atoms of hydrogen and other smaller clumps of hydrogen at an ever more rapid rate. In time, the sphere of the hydrogen atoms, now the core of an embryo star, will become immense and it will begin to rotate on its axis and to orbit around the center of its "mother cloud," the galaxy. As this happens, it will also move out of its place in the spiral arm of the galaxy into the space between the arms. When the move out of its "birthplace" occurred, other things also happened—for embryo planets also were forming and the structure of the Solar System began to take shape. All of this happened in accord with the Laws of Nature and the design that had been set in place.

When the core of hydrogen atoms reached the critical size and the force of the gravitational field that become powerful enough, an embryo star became a full-fledged star. The outer regions of the sphere began to be pulled inward toward the center of the cloud, causing a contraction in the size of the sphere. As this happened, the density of the core increased greatly and the temperature climbed rapidly. As the process continued, the temperature was raised to about 20,000,000°F. At that point, the hydrogen in the core ignited and the atoms of hydrogen began the fusion process which resulted in the creation of helium—the second element of the Universe's chart of elements. Great amounts of heat = energy were developed in this atomic reaction which in time would radiate out into the space of the Solar System. When this ignition happened, the Star, our Sun, was born. "The atoms in the interior of the cloud, drawn to the center by the force of gravity, piled up to form a dense, hot mass. The temperature at the center of the cloud climbed; nuclear reactions flared up, and the sun was born" (Jastrow, *Sun*, p. 54).

To make sure we are getting the right story from the scientific community relative to the birth of the solar system by happenstance, we turn to Ditfurth: "The story of our sun began between 6 and 8 billion years ago when a gigantic cloud off interstellar matter gradually began to contract. This finely diffused cloud was composed of hydrogen atoms with a small mixture of heavier elements; initially it was hundreds of times larger than our present solar system. (This would give it a diameter of trillions of miles for our present Solar System is calculated to be 10 billion miles in diameter.) The contraction of the cloud resulted from the mutual gravitational attraction of the hydrogen atoms. As the atoms contracted toward their common center of gravity, the center of the cloud continually grew more dense. The increasingly massive cloud core gradually intensified the gravitational pull it exerted on the outlying atoms. As the core increased its attraction, the cloud contracted faster and faster. This process continued over an immense span of time. As they contracted toward their common center of gravity, the hydrogen atoms tended to graze or bump into one another. From the very beginning, the cloud had been undergoing a second form of motion; while continuing to shrink, the entire formation was also turning on its axis like a giant carousel.

"As the giant sphere contracted, it steadily grew more compressed. This growing density built up enormous temperatures and pressures at the core. At last the tremendous heat and pressure set off nuclear fusion reactions in the core. These atomic reactions continued to release huge quantities of energy. With the development of atomic energy, the gaseous sphere became a genuine star" (Ditfurth, p. 60).

Thus, the scientists agree on how a star develops. Their findings vary in details but the basic story has been set. In the matter of the development of the Solar System, the same kind of agreement among the scientists appears to prevail. "The outer regions of the cloud—cooler and less dense—gave birth to the planets. In the first step of the birth, *atoms collided and stuck together to form microscopic grains of solid matter.* Silicon and oxygen joined with aluminum, iron and other substances to make small fragments of rock. Iron condensed out separately to form tiny, dully glinting grains of pure metal. (At no time does the scientist explain how these elements were formed or how they got into the cloud which they said was composed 75% of hydrogen and 25% of helium!). As time passed and the cloud cooled further, water molecules froze into crystals of ice. Each ice crystal and grain of rock or iron circles in its own orbit around the sun like a miniature planet. The earth itself did not yet exist.

"This was the solar system in its earliest years: a vast cloud of gases held together by gravity; at the center, the blinding yellow globe of the young sun; and, surrounding the sun, an irridescent halo of ice crystals and rocky grains, drifting slowly in planetary orbits" (Jastrow, *Sun*, pp. 54f).

Jastrow continues his account, telling how our planet Earth

likely resulted from "random collisions occurring now and then between neighboring particles in the course of their circling motion," until, after millions of years the small fragments grew into larger ones and as they became larger, their gravitational attraction increased the tempo of growth. This was also happening with "other microcosmic grains" and when they grew large enough "they quickly swept up all the materials in the space around them, and developed into full sized planets in a short time". He then states, "At the end, most of the material in the solar system was gathered into the nine planets, and only a few fragments and atoms of gas remained in the space between."

To check out this account, we will again turn to Ditfurth. "As the contracting cloud grew smaller, it began to spin faster and faster. That is, its increasing compression also increased its speed of rotation. . . . Eventually, a fragment of the now disc-like cloud broke away. Although this shred comprised less than one percent of the cloud's total mass, it carried away about ninety-nine percent of its impulse of rotation. . . . In time the shred of matter which had broken away from the contracting cloud formed the nine planets. The planets retained the impulse of rotation that had been torn away from the cloud. . . . After the cloud fragment had broken away from the disk, the cloud's powerful gravitational field overcame the remains of the centrifugal force which had tended to flatten the formation. That is, the attraction of the core caused the cloud to resume the shape of a sphere. This sphere became our sun. Almost all of the original cloud went into the formation of the sun, which now accounts for almost 99.9 percent of the total mass of our solar system. Slightly more than 0.1 percent of the total mass makes up all the planets, moons, meteors, comets and interstellar dust. In terms of sheer mass, the sun is not merely the center of our solar system, it *is* our solar system. The nine planets (including our earth) and the rest of the system weigh virtually nothing in comparison" (Ditfurth, pp. 61f).

It will be noted that the two accounts are not exactly the same. Jastrow has the planets growing from microscopic grains into their present forms while Ditfurth indicates that a shred of matter spun off the swiftly rotating cloud and in time it formed the nine planets and all the other bodies in the solar system, such as moons, asteroids, comets, etc. Both have the planets forming out of the original Solar System cloud, both indicate that the total mass of the material that went into these formations was very small in comparison to the mass of the sun. Both accounts indicate that only the material that was in the original solar system cloud could have been incorporated into the formation of the planets and the other bodies we find in the solar system.

In order that we might keep a proper perspective on this very important point in the story of the birth of the Solar System, we must digress for a bit to remind ourselves about the size of the original Solar System cloud, the size of the galaxy out of which it came and the size of the Universe, of which the Galaxy is just a small part. The original Solar System cloud, as we have indicated earlier, may have had a diameter of trillions of miles! Yet, it was just one cloud within the Milky Way Galaxy that developed a star and a solar system. We know that our galaxy has more than 100 billion similar stars—ours is called an average-sized star. Thus, more than a hundred billion clouds, each with diameters of trillions of miles, must have emerged from the truly gigantic galaxy cloud, which is our Milky Way. We do not know how many of these stars developed solar systems but possibly many of them did. The Milky Way Galaxy was not diminished by the star-formation process—it's mass was simply taking new forms. Our Solar System, now with a diameter of ten billion miles, is in orbit around the center of the Galaxy. Traveling at the rate of about 162 miles per second, this huge sphere takes about 250,000,000 years to make one orbit! It is believed that some stars take 400,000,000 years to make one orbit and that some may travel at a speed of 300 miles per second.

Besides the star systems in orbit within the Galaxy, there is also another group of very old stars, dating back to the beginning of the Galaxy. These are called globular clusters. About 119 have been identified in the Milky Way Galaxy. They form a sphere which is believed to be representative of the shape of the original galaxy cloud that separated out from the Creation cloud some time after the Big Bang. The diagram of the Milky Way Galaxy shows the globular cluster pattern superimposed on the Galaxy. The globular clusters do not follow the same kind of orbit that the solar system has developed. They appear to follow an orbital pattern that was developed long before the star system in the Galaxy was born. there are billions of stars in the cluster formations—each cluster having millions of stars. Other Galaxies have the same kind of "old-star" formations—the Andromeda Galaxy, which is our closest galactic neighbor—even though it is 12 million trillion miles away—has been studied and more than 200 globular clusters have been identified. In the following chapter, when we discuss the problems that have arisen with respect to the theory of element-formation, we will take another look at the mysterious globular clusters.

It is calculated that there are in the Universe hundreds of billions of galaxies. Thus, if each of these galaxies developed star systems similar to that found in the Milky Way—and it would appear that they likely have—our language is not big enough to find words to express the numbers of stars involved nor are our minds able to comprehend. It could be that the figures involved in this kind of Universe are so "mind-boggling" that the impossibility of the human mind to comprehend how a Creator-God can have such an operation under control leads to the agnostic position! If this is so, we must sit back for a moment and realize that the created ones must in no way attempt to put a "measure" on the Creator-God, who is, and must be, of a Kind beyond all human understanding.

The scenario of the development of the Solar System, as it has been told by scientific evolutionists, has serious flaws. There are questions that must be faced and if possible, answers should be provided. There is no problem with the concept that a huge hydrogen cloud in a spiral arm of the Milky Way Galaxy moved out into open space between the arms and began to contract, developing at its center a core of intensely hot hydrogen atoms. There is no problem with the concept that when the temperature at the core of the cloud had increased to about 20,000,000°F, the hydrogen ignited, fusing hydrogen nuclei into helium nuclei. There is no question about the establishment of the nuclei furnace at the core of the new Sun, using hydrogen as fuel. There is no question that the nuclear reactions produced huge quantities of energy. The problems begin to surface when we realize that the Sun, in its current phase, is not able to produce elements heavier than helium, next to hydrogen, the lightest of all elements. There are 103 elements possible in our Universe—except for the unstable elements that have been created in the laboratories since scientists have learned how to split the atom. All 103 of these natural elements are found on Planet Earth! The question that needs an answer from the scientists is this: as the original cloud that contracted to form the solar system apparently did not contain all of the 103 elements and as the Sun is not able at this stage in its life to create elements heavier than helium, where did the other elements come from?

The question is of such importance that we must restate it. The scientific position is that while the sun-formation process was going on in the center of the Solar System Cloud, a tiny part of the contracting mass split off of the spinning embryo sun. Ditfurth calls it a "shred" of matter—but it could only consist of hydrogen and helium atoms plus its share of the "small admixture of heavier elements" which the scientists have ascribed to the original Cloud. Note: The entire Solar System cloud had only a small mixture of heavier elements,—which means that the shred could not have had more than its tiny share. Or, as Jastrow claims, hydrogen began to clump up into small grains and these, in orbit around the center of the cloud, began to attract other atoms of hydrogen and other elements that may have been caught by the gravitational pull of the cloud while it was still in the spiral arm of the Milky Way Galaxy. However they got started, the sense of the scientific position is that these microscopic grains, or bits of the shred of matter, developed into what we now know as the Planets and the other bodies of the Solar System. It should be remembered that at the time of their formation, these grains of matter were not organized bodies, just portions of the contents of the contracting cloud. Jastrow puts it well: the hydrogen atoms, together with atoms of other elements held in the cloud by its gravitational force, "collided and stuck together to form small grains of solid matter." These in time, by gravitational attraction, grew into the planetary system. As the system developed out of the same cloud as the Sun—the small "mixture of heavier elements" would have been divided proportionately among the bodies of the Solar System—unless some other Force intervened!

By now, a number of questions have been raised. How did it happen that out of the huge pre-solar system cloud with a diameter of trillions of miles, only nine microscopic grains of matter developed into planets? Why was not the composition off these newly formed bodies identical, or nearly so, as the contents of the Cloud should have been relatively uniform throughout? The fact that the few bodies of matter that did develop were located in different parts off the Cloud and thus at different distances from the gravitational center is a part of the answer but cannot be the whole of it. Carl Sagan makes an interesting comment: "We still do not really know why there are only nine planets, more or less, and why they have the relative distances from the Sun that they do" (*Cosmos*, p. 61).

The most important question, however, is: How did it happen that only one of these nine grains or pieces of the shred was able to develop to maturity having all 103 natural elements? The Planet Earth is that one! Was this just happenstance? In accord with the principles laid down by the scientific evolutionists, all of the development of the Solar System must have taken place "In the random motions" of the atoms in the Solar System Cloud. This question needs to be faced—even if the answer cannot be found. Careful reading of the accounts of the scientists on this phase off the development of the Solar System fails to reveal answers to the questions that have been raised. Not only did the Planet Earth obtain all off the elements possible in the Universe, but in substantial quantities, or, if not substantial, what is even more interesting, in the amounts necessary for the development of and the support of life as we know it. According to the statements of the scientists, these elements could not have been developed in the core of the Sun. The theory that earlier stars exploded, throwing heavier elements into space could account for whatever elements were in the Solar System Cloud but that would represent only a relatively small number of the 103 elements. This theory has other flaws which we will discuss when we come further on in the story of the life of a star. Thus, the mystery deepens. Are we faced with with the possibility that the Creator-God intervened with a "Special Creative Act"? at the time of the formation of the Solar System—giving it all the elements needed? While the question remains unanswered—at least for a time—we must move on to look more carefully at the life of a Star as told by the scientific community.

Before we do this, we call attention to an interesting paragraph in Ditfurth's *Children of the Universe*, where he deals with the effect of the moon on the Earth. It illustrates the inability of the scientific writer to face the fact that no explanation has been given as to how the Earth has all of the 103 elements. "Scientists still dispute the question of whether the earth was formed from a cloud of gas or a fine cloud cosmic dust. The sun formed from a gaseous cloud but contains only very small amount of such heavy metals as are found on earth. Scientists have not yet discovered the reason for this difference in composition between the earth and the sun. In any case, the heavy-metal content of the earth argues against its having formed from a cloud of gas. If the earth formed from cosmic dust, then the elements now composing this planet must have been present from the very beginning" (p. 157).

That is exactly the point! The Sun does not have the 103 elements—it is almost all hydrogen and helium—yet it formed out of the same Solar System cloud from which also the planetary system was formed. As the Sun contains 99.9% of the mass of the Solar System it was and is its gravitational center. Thus the tiny Planet Earth could not have competed with the Sun in attracting to itself the 103 natural elements. This would also argue against Ditfurth's statement that a "shred of matter" broke off from the condensing core of the cloud to form the planets. In fact, the whole Solar System cloud was described as being a gaseous cloud of hydrogen—with little mention made of many elements being present. Ditfurth's claim that "almost all of the original cloud went into the formation of the sun" means that what was left went into the formation of the planets, moons, asteroids, dust and gas, etc. and the elements as we find them on the earth were not there! You may be sure that Ditfurth was well aware of the weakness of the position of the scientific evolutionists on this point. If there had been any solid evidence to show the presence of all the elements in the Solar System cloud as it separated out from the galactic cloud, he would have stated this clearly. Paul Davies, in *The Runaway Universe* (p. 38), provides information that touches this point. "It has long been known that the relative abundance of the elements on the Earth is in no way typical of the rest of the Universe." In other words, to have the elements on Earth in such relative abundance is atypical of the Universe—it is unusual, unexpected, not understood. Davies goes on: "The Earth, with its great quantities of iron, nickel, oxygen, copper, and so on, is really just a speck of contaminent, a concentration of all the rarest substances in one place and quite untypical of the cosmic abundance." Thus, we must conclude that the presence of the 103 natural elements on the Planet Earth cannot be accounted for from a scientific basis. We suggest that the mystery could be resolved by allowing for a purposeful intervention on the part of the Creator-God at the time when the Solar System was set aside to give birth to the Sun, the planets and life on Planet Earth.

Not forgetting that we left unanswered some very important questions about the origin and composition of the planets that are in orbit around the Sun, we must follow the development of the Sun itself. As has been said earlier, the core of the huge hydrogen cloud that separated out from a spiral arm of the Milky Way Galaxy, became larger and larger and more densely packed as the contraction process continued. The temperature and pressure at the center climbed steadily. When the critical temperature had been attained—15 to 20,000,000 degrees F, the nuclear fusion process began and the core of the newly born sun became a mighty furnace. When this happened, a new force became evident. The inward pressure and heat built up by the process of contraction of the Solar System cloud under the influence of the gravitational force of the core, were now countered by the outward pressure formed by the heat of the nuclear reactions at the core. The first type of pressure was powered by the gravitational field force pulling the cloud inward to the center of the new Sun. Now, within the core of this Sun, the new force created by the nuclear reactions pushed outward and the contraction process was stopped. Actually, the contraction process was reversed by this new force and the core began to expand. When this happened, the temperature dropped and the atomic fires went out. This released the outward pressure and the contraction due to the power of gravity began again. This brought the temperature back to the atomic ignition point, again reversing the action. The new Sun was in an unstable condition in this period. The alternating contraction-heating, cooling-expansion process caused a kind of pulsating action to occur. This condition continued for possibly many millions of years before an equilibrium was reached and the pressure of expansion equaled the pressure of contraction bringing the Sun into a period of stability with the nuclear furnace still burning. Since the Sun has remained stable for more than four billion years and so far has consumed only about half of its hydrogen supply, the scientists expect it to continue its stability for a few billion years longer! Astronomers are observing that in the areas within the Galaxy where they believe new star-formation is taking place, the same

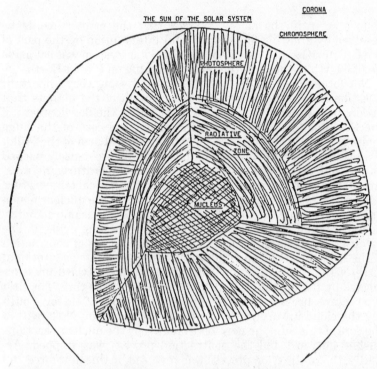

THE SUN OF THE SOLAR SYSTEM

The Sun, 99.9% of the mass of the Solar System, is about 70% hydrogen and 30% helium. These gases exist under great pressure and temperature—mostly in the form of their nuclei—as their electrons have been stripped from them. The Sun, however, is 864,000 miles in diameter; it has a core about 218,000 miles in diameter, with a temperature of 15, to 20,000,000°F and a pressure of 200 biillion tons per square inch! It is in the core that the nuclear fusion process occurs - the "burning" of hydrogen - producing huge amounts of heat-energy and what for the time being is a "waste product" - helium. 657,000,000 metric tons of hydrogen are converted into 652,500,000 metric tons of helium each second. The 4,500,000 ton difference represents the "heat of the Sun" as it is formed into gamma rays, x-rays and neutrinos. Energy quanta, traveling at a little less than the speed of light, requires 20,000 years to move from the core of the Sun to the surface. There, as a photon, a light ray travels to the Planet Earth in eight minutes, bringing the energy necessary for the functioning of the Planet and for life as we know it.

kind of pulsations among these new "young" stars is occuring. It evidently is a normal phase in the development of a star.

The studies which have been made in these recent years have brought the scientists into possession of information about the Sun that is truly astonishing. The Sun has a diameter of 864,000 miles; it weighs 330,000 times as much as the Planet Earth. It is 92,900,000 miles from the Earth. Each square centimeter of its surface radiates 1,500 calories of heat-energy per second. The pressure at the Sun's core equals more than 200 billion tons per square inch; the temperature at the core is is about 15,000,000°C. The matter at the Sun's core is 12 times as heavy as lead but it is fluid, a plasma. The hydrogen and helium atoms are ionized, that is, they have lost their electrons, only the nuclei are there—densely packed in. About 70% of the Sun is hydrogen and 30% is helium. There are trace amounts of heavier elements.

The active zone of the Sun's core where the atomic fusions occur, has a diameter of 218,000 miles. As the hydrogen nuclei collide and form nuclei of helium, heat-energy is released in the process as the helium nuclei weigh somewhat less than the hydrogen nuclei that are involved. This small difference in mass translates into pure energy, as mass-energy cannot be lost—it can only be transformed. Thus hydrogen is constantly being used up and the amount of residual helium increases. The helium, in turn, is inert and does not react to any of the forces of heat or pressure in the core. Thus the helium has been called the "ash-heap" in the core of the Sun. Much later in time—as we shall see—the helium will become the atomic fuel when the hydrogen has been exhausted. This will not happen until long after the Solar System has been destroyed and the Sun, on the way to become a White Dwarf, has begun a new contraction period. This will increase the temperature of the core to about 100,000,000 degrees. Then the helium will ignite and the atomic furnace will be many times hotter than when the fuel was hydrogen.

Each second, 657,000,000 metric tons of hydrogen are converted into 652,500,000 metric tons of helium. One metric ton equals 2,200 pounds. The difference, 4,500,000 tons, provides the mass-energy which we think of as the heat of the Sun, neutrinos and energy in the form of gamma rays and X-rays. "The sun comprises virtually our entire solar system. But its very bulk is essential to our lives. If the sun were not so large, it could not build up the enormous temperatures and pressures in its core, that is, it could not generate atomic reactions. Moreover, the sun's bulk acts as a shield to protect us from the energy at it's core. If this energy were to strike the earth in its original form—gamma rays and X-rays—none of us would be alive today. But because our sun is so vast, the earth and its fellow planets are bathed in light and warmth rather than in deadly radiation" (Ditfurth, p. 70). This constitutes another evidence of plan and design built into our Universe and our Solar System. The amount of heat generated by the Sun is needed to keep the Solar System going—but the Planet Earth needs protection from its intensity—and gets it. As we will see in our next chapter, the protections built around the Planet Earth are unexplainable on any other ground but design with purpose.

Ditfurth goes on to tell us that each quantum of energy travels through about 373,000 miles of solar matter to reach the surface of the Sun. Even though the energy quanta travel at almost the speed of light, they take about 20,000 years to move from the core of the Sun to the surface. This is so because they cannot pass through in a straight line but bounce in a zig-zag pattern through the atomic nuclei, being absorbed and released by the hydrogen atoms with which they collide. "The light (energy) which has finally worked its way to the sun's surface arrives on earth only eight minutes later; but by now this light is the "ghost" of its former self. Yet this dim reflection of past glory *provides the right amount of light and heat to nurture the frail creatures of earth*" (p. 71, emphasis mine). The scientist apparently overlooks the obvious—preparations had been made for a planet to be "home" for life as we know it!

Within the Solar System there are nine huge spheres moving in clearly defined orbits around the Sun, held in place by the tremendous gravitational attraction of the Sun. The planets are also influenced by the gravitational effects each has on the others. The last three planets, Uranus, Neptune and Pluto, were discovered in the mathematical studies which were made by scientists seeking to explain the perturbations in the orbits of the known planets. It appeared to the mathematicians that the disturbance in the orbit of Saturn could only be accounted for if there were another planet farther out. Their calculations revealed the Planet Uranus. Following that discovery and using the same technique, the Planet Neptune was located for it was disturbing the orbit of Uranus. Finally, the small Planet Pluto, was found. Tiny as it is in comparison to the huge planets, it still causes a perturbation in the orbit of Neptune. Our diagram below, shows the planetary system—not drawn to scale. Thus, in a mysterious manner, the planets do constitute a "system," not only moving about the Sun in an orderly pattern but having an effect on each other—all the while moving with the Sun in its giant orbit around the Milky Way Galaxy.

PLUTO
NEPTUNE
URANUS
SATURN
JUPITER
ASTEROIDS
MARS
EARTH
VENUS
MERCURY
SUN

The origin of the planets is still a mystery. Many theories have been advanced, such as an explosion of a "companion star" of the Sun, or the "near miss" of a collision of another star with the Sun in which a part of the Sun is supposed to have been torn away—the shred to have broken up in time to form the planets. It appears now that the most widely accepted position of the scientists is that the planets came into existence at the same time as the Sun itself, being born out of the huge Solar System cloud in much the same manner as the Sun.

The mystery of the origin of the planets deepens when we look at them individually. In spite of the fact that they all appear to have developed out of the substance of the Solar System cloud, each is different. They are similar in that they all are in orbit around the Sun, all but one turn on their axes, they all have mass which is controlled by the gravitational force of the Sun. They also exert an influence on each other. The sum of the gravitational forces of the nine planets must also exert an influence on the orbit of the Sun about the gravitational center of the Galaxy. The similarities appear to be primarily in the area of function. Planets are dissimilar when compared on the basis of size, substance and in a number of items such as temperature, atmosphere, density and life possibilities. We will look briefly at each one.

Mercury orbits the Sun at a distance of 35,960,000 miles. It has a mean diameter of 3,100 miles and its average speed of rotation is 107,900 miles per hour. It circles the Sun once in 88 days and so defines its year. The planet does not rotate on its axis and so does not have a day-night sequence. Thus the same side of the planet is always facing the Sun which means that it is very hot on that side—an average of 350°C—but reaching much higher temperature when the planet's orbit brings it closer to the Sun. The "night side" temperatures average −170°C but comes close to absolute zero or −273°C when the planet is farthest out from the Sun. There is no atmosphere and thus there is no life of any kind on Mercury. It has no moon.

Venus orbits the Sun at a distance of 67,200,000 miles. It has a mean diameter of 7,700 miles—close to that of the Earth at 7,920 miles. Its average speed of rotation is 79,000 miles per hour. It circles the Sun once in 225 days, which defines its year. It revolves slowly on its axis, possibly only once in 30 earth days. At ground level the temperature is about 450°C to 480°C. It has a heavy atmosphere, much thicker and denser than that of Earth, It consists of carbon dioxide, 97%, with just a trace of water vapor. "With its density one tenth that of water and its temperature 100°C hotter than molten lead, the physical conditions of its atmosphere near the surface of this planet are scarcely imaginable to us. . . . The clouds of Venus are formed of very minute drops of sulphuric acid" (Maffei, p. 39). There is no possibility of life on Venus.

Earth orbits the Sun at a distance of 92,900,000 miles. It has a mean diameter of 7,920 miles. It circles the Sun once in 365.25 days, which defines its year. Its average speed of rotation is 67,000 miles per hour. Its day is 23.56 hours in length. It has 1 moon—230,000 miles distant. Its temperature averages about 15°C which makes it ideal for life. It has an atmosphere composed of nitrogen 78+%, Oxygen 20.+%, water vapor and very small amounts of rare gases. There is a huge amount of water on the planet in the form of oceans, lakes and streams. There is also a large amount of water in the soil and atmosphere. All of these features are remarkable in that they are necessary for the support of life. No other planet is like the Earth in these respects and no other supports life.

Mars orbits the Sun at a distance of 141,600,000 miles. It has a mean diameter of 4140 miles. Its average speed of rotation is 54,200 miles per hour. It circles the Sun once in 667 days, defining its year. Its day is 24 hours 37 minutes in length. It has two moons, about 6,000 and 14,000 miles distant. Its mass is about 0.11 times that of Earth. Its density is 3.9 times that of water. Its axis is inclined to a 23°59' angle which is slightly more than the inclination of the Earth. This gives it seasons somewhat similar to that of Earth but longer because Mars is farther from the Sun. The planet has mountains, valleys and plateaus, numerous extinct volcanos and some that may still be active. It has craters like those on Mercury and the Moon of Earth. The atmosphere is very rarefied, the pressure at ground level being about what Earth would have at some 20,000 feet of altitude. The atmosphere is almost all carbon dioxide—95%, nitrogen—3% and argon 2% with some trace gases. Water vapor is rare—only 0.01%. At the two poles there are white caps of carbonic acid ice with some water ice. It is believed that there is water ice also in the high mountains. The temperature is cold—a high of 15°C in midsummer to $-100°C$ in winter. No indication of life has been found by the space probes that have been made even though Mars is famous in science-fiction stories about an advanced civilization with a grand canal system. It does appear that at one time there may have been streams of water which now have evaporated into the atmosphere and beyond.

The missing planet: Astronomers have long believed that there should have been a planet between Mars and Jupiter. Instead, a band of asteroids has been found orbiting the Sun in a region where the planet should have been located. They appear to be composed of rocky substances, the largest being about 500 miles in diameter while most of them are very much smaller. It has been calculated that there are at least 22,000,000 of these small bodies and some astronomers believe that they may number in the billions. If it was a Planet, it must have exploded a long time ago. None of the asteroids has an atmosphere.

Jupiter orbits the Sun at a distance of 483,300,000 miles. It has a mean diameter of 36,800 miles, making it almost 11 times as large as the Earth. It is the largest of the Planets. Its average speed of rotation is 29,400 miles per hour It weighs 313 times as much as the Earth. Jupiter has 12 moons in orbit around it. The atmosphere is about 500–700 miles thick and is composed of hydrogen, helium, amonia, methane and water vapor. There is no solid surface on the Planet. The surface consists of an ocean of molecular hydrogen liquid, possibly 15,000 miles deep! The temperature at the bottom of the ocean is 11,000°C with a pressure of 3 million atmospheres. Below this depth, the hydrogen passes into a metalic state and at a still greater depth, the hydrogen under much greater pressure dissolves into liquid hydrogen atoms. The core of the planet has a radius of possibly 2000 miles and is composed of iron and silicates at a temperature of about 30,000°C. Naturally, there is no life on this planet. Jupiter, however, has a heat source within it for it emits twice as much heat as it receives from the Sun. Some astronomers believe that Jupiter almost became a star. No other planet is like it in this respect.

Saturn orbits the Sun at a distance of 886,200,000 miles. It has a mean diameter of 71,500 miles. Its average speed of rotation is 21,700 miles per hour. It circles the Sun once in 29.5 Earth years. Its day is 10 hours and 38 minutes in length. It is 94.15 times as heavy as Earth. Saturn has nine moons. The physical structure of Saturn, including its atmosphere, is almost identical to that of Jupiter. Saturn is famous for the beautiful rings of meteoric particles which revolve in distinct orbits about the planet. There is no possibility of life on Saturn.

Uranus orbits the Sun at a distance of 1,783,000,000 miles. It has a mean diameter of 32,000 miles. Its average speed of rotation is 15,310 miles per hour. It circles the Sun once in 84 Earth years. Its day is 10 hours and 45 minutes in length. It has 5 moons. Uranus is 14.5 times as heavy a Earth. It is so far out that even the most powerful telescopes can reveal little of its physical characteristics. It is believed, however, that it is similar to that of Jupiter and Saturn. There is no possibility of life on Uranus.

Neptune orbits the Sun at a distance of 2,794,000,000 miles. It has a mean diameter of 27,000 miles. Its average speed of rotation is 12,240 miles per hour. Neptune circles the Sun once in 164.75 Earth years. Its day is 15 hours and 48 minutes in length. It has 2 moons. It is believed to have a structure and atmosphere similar to that of the other "far-out" planets. The temperature must be less than $-200°C$. No life is possible on Neptune.

Pluto orbits the Sun at a distance of 3,670,000,000 miles. Its average speed of rotation is 10,800 miles per hour. Pluto is believed to have a mean diameter of about 3,600 miles. It circles the Sun once in 243.5 Earth years. Its day is believed to be about 6.5 Earth days. Its mass is possibly only 0.11 times that of Earth. It is believed that because of its intense cold, $-230°C$. any atmosphere must be frozen. It is also believed that there is a frozen surface of ice and snow. No life is possible. Some astronomers think that Pluto might not have originated in the Solar System cloud but that it is a celestial body from outside the solar system that got caught in the gravitational pull of the Sun. Pluto is so greatly different from the other far-out planets, being much smaller and having a solid surface of ice and snow—not necessarily water—makes it appear quite possible that it is not a true planet of the solar system.

Having taken this brief look at the planets, there are interesting observations to be made. The "inner" four planets are small in comparison to the "outer" four—excluding Pluto. They have solid surfaces and thus contain more of the heavier elements. The bulk of the outer planets consists of gases of hydrogen, helium and methane, with very little water vapor. Their interiors are hydrogen in liquid and metallic forms, plus free hydrogen atoms—with a nucleus of iron and silicates at a high temperature. Jupiter, the largest of the planets, produces more heat than it receives from the Sun but does not support atomic reactions that could have made it a star.

The planet Earth has marked differences from all of the others. It has a temperate climate. It has an atmosphere made up of nitrogen, oxygen, water vapor and a few rare gases. The others have either carbon dioxide gas (Venus and Mars) or hydrogen-helium-methane gases (Jupiter, Saturn, Uranus, Neptune). Mercury has no atmosphere and if Pluto has an atmosphere, it would be frozen. Earth has an inclination of $23½°$ which gives it four seasons. The surface has a plentiful supply of soil and water which none of the others have. It also has ample supplies of oxygen, nitrogen and carbon, elements essential to the support of life, both plant and animal. Plant life on Earth constantly supplies the atmosphere with free oxygen which is necessary for animal life. In turn, animal life provides carbon dioxide to the atmosphere which plant life must have to exist. Thus, each form of life is dependent upon the other for survival. Only the Planet Earth supports life. All the evidence indicates that none of the other planets could do that now or ever has done it in the past.

The unanswered questions we faced earlier in the chapter not only remain unanswered but new ones of a similar nature have arisen. Only one planet has a temperate climate, an ample supply of water, oxygen, nitrogen, carbon and other elements essential to life as we know it. Did this happen by chance or was it the result of design with a purpose?

The unanswered questions must be faced by the scientific community. We have given a possible answer, which for the time being we will allow to stand—that possibly the Creator-God intervened to provide the Planet Earth with the ingredients necessary to sustain the life He purposed to establish. The scientific community has failed to give an answer other than resorting to the concept that somehow, the passage of huge amounts of time allowed the Solar System cloud to condense and form the Sun, the planets and all else we have found in our present Solar System, including life, plant and animal—which, of course, is no answer at all to the questions we are facing.

CHAPTER VII
THE PLANET EARTH, PART I: THE MYSTERY OF THE ELEMENTS

Whether Planet Earth and the Solar System in which it is located is the center of the Universe or simply a small planet in an average sized solar system in an average sized galaxy is not of the greatest significance. What is important is that it is where we live! We need to know as much as possible about it. There are mysteries about Planet Earth that appear to be unexplainable and that make the Solar System and the Planet very special, possibly unique in the catalogs of bodies in space. Maybe there are other solar systems with orbiting planets that are similar, but no one *knows* this for sure. If other similar systems should be identified, the mysteries would not be resolved but only deepened. One thing we do know. Planet Earth is our home. We live on the Earth; we are nourished and protected by it. If it were not for the mysteries, functioning as they do, life would not be possible on Earth, we would not be here!

We need to look at these mysteries so that we can appreciate their significance even though we may not be able to understand and explain them. Knowing of them, realizing their value to us and accepting them as facts of life which have served our Solar System and planet for billions of years, will make us more comfortable, even though the mysteries remain. We will look first at the scientist's theory of the method by which the development of the 103 elements on Planet Earth may have occurred in the fiery nuclear furnaces at the core of the stars. We do not mean to suggest that elements are not formed in this manner. Rather, the story as it is told is not complete and in places appears to be self contradictory, as we shall see.

The second mystery that we will probe in the following chapter, consists of three protective devices that shield the Planet from deadly cosmic radiation and thus help make life on Earth possible. These mysteries all have their origin in the Sun, the star of our Solar System.

We have questions relative to the theory of how the elements were created and as to how all 103 natural elements came to be on Planet Earth. These questions, in our opinion, have not been dealt with adequately in the writings of the scientists we have used as references. In handling the subject they tend to become quite vague, defensive and at times resort to what appears to be the use of camouflage to cover the blank areas in their explanations. It appears that the scientist too readily uses the notion that huge amounts of "undirected" time can accomplish miracles. Look again at the structure of atoms of elements as described in Chapter Four. The very precise building up of elements from hydrogen on through uranium, with the orderly yet complicated regulations with respect to the development of the electron shells, simply does not fit with the concept of random explosions of Supernova Stars. The contraction of the Solar System cloud, under the pressure of gravitational force produced friction-caused heat and pressure which in time developed the Sun at the core of the Cloud. The Sun has the power to develop helium by the fusing of hydrogen nuclei but no other element in its present phase of life. Thus, it could not have supplied the Solar System or the planet Earth with the other natural elements.

As all the planets, as well as the Sun, were formed at the same time from the contents of the Solar System Cloud, and as the Sun which contains 99.9% of the mass of the Solar System does not contain the heavier elements, how can we account for these elements as they are found on Planet Earth? "Initially, the Sun was an even mixture throughout, 75% hydrogen and 25% helium (with an almost immeasurably small smattering of heavier elements)" (Gribbin, *Sun*, p. 64). This mixture of hydrogen and helium gas plus the small admixture of heavier elements is exactly how the original Solar System Cloud was described. Note that the scientists never say that all 103 elements were present in the Solar System Cloud. Nevertheless, the fact of the existence of these elements on Earth is real—they are here and we use them! Thus we must conclude the 103 elements were present in the Solar System Cloud when it was formed. How they got into the Cloud that formed in the Milky Way Galaxy is an unanswered question.

Planet Earth and the other three inner planets were assembled out of the contents of the Solar System cloud. The only way to account for the composition of Planet Earth—and the other inner planets—is to accept the concept that in the formation period there was a "holding action" preventing the heavier elements from being drawn to the Sun. As the gravitational force of the newly born Sun increased, the heavier elements were separated out, held in orbit in approximately their present location in the terrestrial planets as they were forming. The same kind of action was happening in the development of the outer planets composed basically of hydrogen, helium, methane and other gases except that in the cores some of the very heavy elements are found—defying, in a sense, the Laws of Nature in that they were not being pulled into the Sun by the force of gravity. How each of the planets achieved their growth and secured their present orbits is not clear.

With matter not acting in accord with natural law, the "intervention" of a Superior Force—a Creator-God—at this point would clear up the problem that is apparent. Between the inner four heavier-element planets and the lighter-element outer four (excluding Pluto), we find a huge band of heavier-element matter all in a regular orbit about the Sun. It is believed by many astronomers that these pieces of matter—some are 500 miles in diameter while others are microscopic in size—represent debris from an exploding planet. This may well be so—but it could also be that this so-called asteroid belt could be remnants of heavier-element particles that somehow never became attached to a planet during the period of planet-formation and are now caught in an orbit which will remain fixed as long as the Solar System exists.

It is one thing to accept as fact that all of the heavier elements contained in the Solar System cloud were found in that one-tenth-of-one-percent of the total mass of the Cloud that was not a part of the body of the Sun. It is another thing to explain how these heavier elements, scattered throughout the enormous space of the Galaxy, came to be in the Solar System cloud, especially in view of the fact that the concentration of heavier elements as found in the Solar System Cloud is in no way typical of their concentration in the Universe. It would be almost inconceivable that the Solar System cloud could have picked up these elements by happenstance and that then they were assembled and put in place on the Planet Earth. By accident? We can only say that it happened because it was in accord with the Laws of Nature as put in place by the Creator-God plus an Intervention Event which held the heavier elements in place as they are now found in the planetary system. One Planet only has all 103 of the possible natural elements because it was in the Grand Design that it should be the home of living plants, animals and humans.

The scientific community has bound itself to the theory that this all came about by happenstance. Thus, we have no answer from the scientists. When, however, we accept the concept that it all happened as the Creator-God worked out his will for the Universe, the parts of the problem fall into place. All of the elements of the Universe have been created—possibly in the manner suggested by the scientists, but also in accord with the Laws of Nature—with an "assist" from time to time by the Creator-God and his workmen. Thus, the elements required for the Planet Earth were in the Solar System cloud; they were sorted out during the formation period of the Solar System in such a manner that the Earth was properly supplied. The orbit of the Planet was fixed in the proper place so that the energy coming from the Sun would be just right for a planet that was to develop life as we know it. The orbits and compositions of the other planets were also fixed in accord with the design and purpose of the Creator-God. Remember, Carl Sagan stated in *Cosmos* (p. 61), "We still do not really know why there are only nine planets, more or less, *and why they have the relative distances from the Sun that they do.*" While the Solar System is almost infinitely small in comparison with the Milky Way Galaxy—and especially so when compared with the total of the Universe—it holds a significant place in the working out of the sustained expressed will of the Creator-God. It may be that there are other Solar Systems like it—we do not know for sure. We do know that we are and that we live on our Planet Earth in our Solar System.

No matter how the elements of the Universe were formed or how they came to be on the Planet Earth, it is necessary for us to be reasonably sure that we understand something about their formation. Because element-formation has been tied explicitly to the life story of the Stars, we need to take a good look at what the scientists tell about the matter. "Once our galaxy had acquired its discus shape, the first generation of stars came into being . . . Once atomic reactions began to occur, the stars were well on their way to producing the rest of the ninety-two elements composing our present world . . . Billions of years passed. Innumerable suns transformed their hydrogen into helium, then changed the helium into carbon, oxygen, and other new elements. Once more, billions of years went by. All the stars were destroyed in enormous explosions. In exploding, they released new elements into space in the form of fine interstellar dust, which gave birth to a new generation of stars. This new generation was enriched by the heavier elements; in turn, it generated still heavier ones, and so the process continued, with the destruction of each star generation creating the raw material for higher development . . . Around 5 billion years ago, cosmic evolution entered its second stage. *New suns were being born which for the first time contained traces of every one of the ninety-two natural elements.* The suns began to form planets from these ninety-two elements. *Of course, the only planets we have actually observed are those in our solar system*" (Ditfurth, p. 286, emphasis mine).

Ditfurth and Gribbin tell basically the same story but with some differences. Ditfurth uses two star types, the average-size star like our Sun, which cannot provide new elements to the Galaxy, and the type which is 1.4 times the mass of the Sun or more. This latter type does produce heavier elements which are blasted into space when the large star explodes at the end of its life—if the accounts of the scientists are correct. Gribbin uses three star types, the average, like our Sun, plus a larger one which can produce a limited number of elements up to iron, and the very large star, possibly eight or more times the mass of our Sun, which picks up the elements from iron on through uranium. These latter types of stars in their explosions become Nova and Supernova stars. It must be said that both men are a bit vague about how it all happens. Their references are always just to the "heavier" elements—they do not speak to the point of the development of all 103 elements specifically. Yet, if we are to have our Universe, Galaxy, Solar System and especially the Planet Earth, all 103 must be accounted for.

"Our world is composed of ninety-two natural elements. These elements might never have come into being—if it were not for the stars! All stars have their personal biography; they live through a definite life cycle. Throughout its life, every star must maintain a balance between two forces, the atomic energy in its core, which presses outward, and the gravitational force which presses in, holding the star together. Eventually, a star uses up all the hydrogen fuel in its core; that is, the star changes all of its hydrogen into helium. For a time the star continues to maintain its stability. Once the hydrogen in the core has been changed into helium, the atomic fire slowly spreads outward from the core, consuming hydrogen in other regions of the star. But, eventually, the atomic fire eats its way near to the surface. Soon the outer layers of the star can no longer exert enough pressure to set off atomic reactions. The atomic furnace burns out" (Ditfurth, p. 274). This sets the stage for the succeeding episodes in the life of a star, which, according to the scientists, leads to the creation of the elements from helium to uranium.

At this point we will use Gribbin, as our guide. At the outset, we must distinguish between three types of stars. While all three go through the same kind of life experience, burning hydrogen and helium to counter the inward pressure of the gravitational force, only the second and third types go on to the point where they discharge newly formed elements into the space of the Galaxy. The first type, like our Sun, while developing some new elements in the late stages of its life as it becomes a White Dwarf, is unable to discharge them into space and thus does not contribute to the element-formation process in a manner that is useful, as far as we know. The size of a star determines how fast it will burn its hydrogen fuel and this determines the length of its life. The larger the star the more energy is required to off-set the inward pressures of the gravitational force. Thus the very large star must burn its hydrogen at a very rapid rate. This means that the larger the star, the shorter will be its life span. This is a critical matter relative to the production of the heavier elements for use by later generations of stars.

The first star type, the one of average size, constitutes the bulk of the stars in the Galaxy. Ninety percent of all stars are in this grouping, many of them being smaller than our Sun. The ten percent, however, account for 25% of the total mass of all stars in the Galaxy which means that they must be very large. The average sized star does not have the mass necessary to develop the heat and pressure that must be present to create the heavier elements. "The heaviest elements a star like the Sun can build up—late in its life time—are carbon and oxygen. As the Sun ages, an inert core of these two elements will be surrounded by a shell in which helium is being 'burned' to make more carbon and oxygen, and by a shell in which hydrogen nuclei are being fused into helium nuclei. These shells move outward through the star with the core growing bigger until the nuclear fuel is exhausted and/or the outer layers are lost in space" (Gribbin, p. 72). Thus, the Sun and all the other "average" stars, do not contribute to the supply of elements in the Universe.

When the nuclear fuel in an average-size star is exhausted, the outward pressure is lost and the power of the gravitational force takes over. As the core of such a star is compressed it finally begins to collapse. When this happens, the outer layers of the star begin to expand as the surface becomes cooler; the star grows larger. The color changes from the yellow of the original star to a red and the star is then called a Red Giant. It is moving on to its death. After several hundreds of millions of years the star will be about 100 times its normal size. In the case of our Sun, this expansion will bring its still hot surface much closer to that of the Earth and life on our Planet will be destroyed. During the Red Giant stage in the life of a star, the helium in the core begins to burn when the heat caused by the gravitational force reaches 100 million degrees. When this happens, the star is again able to counteract the gravitational force until the supply of helium is exhausted. In the final phase of the Red Giant stage, the star will become up to 400 times its original size. In the case of our Sun, its surface will reach out to swallow up the four inner planets.

When the helium is finally gone, the gravitational force again asserts itself and a star like our Sun will collapse into what is called a White Dwarf. The heat produced in the collapse is so great that the small body, now about the size of our Planet Earth, shines brightly, which gives it the name White Dwarf. Ultimately, however, it will cool, becoming first a Red Dwarf and finally a Black Dwarf, just a cinder in the cold space of the Universe. This marks the end of the Solar System—all of its glory will be gone!

The second star type, following Gribbin's model, is much larger than our Sun—at least 1.4 times as large. These stars, because of their great size or mass, must burn their hydrogen fuel more rapidly in order to counter the inward pressure of the gravitational force. Thus they come to the time when the hydrogen is exhausted sooner than was the case with the smaller stars. The burning of helium also begins sooner. This point is made because it indicates that great numbers of star "generations" have likely come and gone—scattering into space any new elements they may have made during the active phase of their life.

Returning, then, to our story of the life of the larger star, when the helium fuel is gone, the star, because of its size, does not move to the White Dwarf stage at once but redevelops high temperatures because of the pressure built up again by the gravitational force which begins the contraction process again. The carbon, now in the core, ignites. This produces outward pressure sufficient to again stop the contractions caused by the effects of gravity. A star of this type can go through the cycle of expanding and contracting many times as it swells and shrinks in a pulsating manner, "as fine adjustments to the balance between gravity and pressure occur in its middle", Gribbin tells us. "These regularly pulsating stars are called Cepheid Variables" (p. 77). "The star is, once again a Red Giant; but it still has the capacity to go through the whole cycle again and again, building up the material in the core through successive phases of pulsation, collapse, and giant-hood until the core is chiefly composed of iron, the most stable element of them all." Then, as there remains no more nuclear fuel to produce heat, a cooling of the star begins. The outer layers of the star collapse toward the center, making the star very hot. "But these outer layers of the star contain plenty of material like carbon, which can still form heavier elements of fusion when the temperature rises, so that in the middle of the collapse, the suddenly heated outward layers literally explode, with a lot of carbon and other elements fused into heavier elements at once. The explosion blasts the outer layers away from the star, *dispersing perhaps 1 percent of the total stellar mass into space*" (Gribbin, emphasis mine).

The energy involved in this explosion causes the whole star to shine very brightly. For this reason astronomers called these "novae"or new stars, even though they were actually old stars about to die! From this point on this star gradually cools off and becomes, after millions of years, the same kind of Black Dwarf which the first star type developed when its end came. This star in its Black Dwarf stage is possibly just a lump of iron, covered by a skin of frozen carbon dioxide. But, it has produced some of the heavier elements which it has blasted into space one percent of the mass of the star) a very small bit in comparison with the hugeness of the space of the Galaxy!

The third star type—and it represents in number only a small percent of the stars of our Galaxy—is composed of very huge stars, possibly eight or more times the size of our Sun. They go through the same routine as the other star types until they come to the point when the core has become iron. Here the fusion process ends. To create elements beyond this point a much higher degree of heat must be attained. Up to this time, new elements were formed by the heat and pressure developed in the fusing of nuclei of the elements being used as nuclear fuel. A different method now takes over. "These are stars with eight or more times the mass of our Sun, or even more massive. With seven or more solar masses of material still pressing down on this core from above, and no more energy available from fusion to hold up the star against gravity, very interesting things begin to happen in the middle of such a star" (Gribbin, p. 79).

Gribbin then goes on to explain how this enormous gravitational force causes the electrons and protons of the atoms of iron to be squeezed together to form neutrons—the mysterious energy particle that was also present in the original Creation-Cloud after the Big Bang. As we indicated when we discussed the formation of the first atom of matter, hydrogen, a neutron can enter the nucleus of an atom without changing the nature of the atom. It does add its weight to the mass of the atom it enters. As was also pointed out, the atom—and this is true of atoms of all elements—is mostly empty space with one or more electrons in orbit around one or more protons. Thus, the neutron can easily slip into a nucleus. Now, in the process of element-formation that is taking place in the core of the super-giant star, the atoms of elements, their electrons and protons having been squeezed together to form neutrons, require much less space—for the neutron is so much smaller. Thus, it is like a huge vacuum is suddenly developed and the center of the star collapses. The upper layers of the star plunge into this vacuum and in doing so release an enormous amount of heat, similar to that developed when a Nova Star exploded but on a far greater scale. We quote Gribbin again: "with gravity now playing the dominant role for the first time in the star's life, iron nuclei in the core are ripped apart and turned back into helium nuclei. As much gravitational energy is put back into the core, from the collapse of the outer layers, as had been stored up by nuclear fusion throughout the entire previous history of the star. Still the energy left over is sufficient to tear the star apart as an energetic shock wave blasts outwards from the inner regions, and for a brief period this one star will shine as bright as all the stars of the Milky Way Galaxy put together, pouring out in a matter of weeks as much energy as it had radiated throughout all the millions of years of its existence up to that point." Up to the development of iron, the element-formation process was run by the heat generated by the *fusion* of atomic nuclei. From iron on up to uranium, the process changes to the *fission* of atomic nuclei, wherein much greater heat is generated by the breakdown of the atomic nuclei.

Out of this unbelievably huge explosion heavier elements are formed and these are scattered out into the space of the galaxy to be used by succeeding generations of stars yet to be born—if the theories of the scientists hold. Explosions of these giant stars produce supernova, as they represent the ultimate in the explosive death of stars. In its death, however, a new star, small but exceedingly dense, is born—the neutron star. It is believed that a supernova explosion in our Galaxy occurs once or twice every hundred years. The Crab Nebula, observed by the Chinese astronomers in 1054 A.D. represents such a supernova. Two others have been observed, in 1572 and 1604. We have just observed Supernova 1987A which exploded February 23, 1987. As the astronomers calculated the exact position of the exploding star they were astonished to find that the position was located at the point where a blue supergiant star used to be! The scientists had always been of the opinion that only red supergiant stars could explode as supernova. So, while the characteristics of the new Supernova 1987A are in line with what was expected, the fact had to be accepted that it was a blue supergiant with a mass close to twenty solar masses that had exploded. There will be intensive study made by the scientific community which now has for the first time in history observed a Supernova come into being—having very sophisticated knowledge and technical aids for use.

The supernova stars are thus the greatest source of heavier elements. The ultimate end of a supernova is much different than that of an average star, such as our Sun, or even as a Nova, as we shall see. We now return from Gribbin to what we called the "story line" of Ditfurth. We should not be surprised to find that there are differences in the accounts because knowledge in this field is changing rapidly. Thus, while some of the details differ, the end result is about the same. We feel that it is important that we repeat, in a sense, the story that has already been told, for it is of great significance to our position—that the element-formation account has not been presented in an adequate form. Possibly it cannot be—yet!

We pick up Ditfurth's story at the point where he tells of the burn-out of hydrogen in the core of a star. At this point he is not describing a specific star—it is the process that he is describing. When the hydrogen fuel has been exhausted, Ditfurth tells us, and the atomic furnace at the core of a star has gone out, the outward pressure developed by the fusion of the hydrogen nuclei is no longer present to counter the inward pressure of the gravitational force. Thus, the process of contraction begins again and as it does, heat is generated as the mass of the core becomes more dense. The contraction begins at the core, but the outer regions of the star which have become cooler, are greatly extended, causing the volume of the star to increase greatly. The color of the star changes from the yellow of the hydrogen sun to a reddish hue. The scientists have called this the Red Giant stage in the life of a star. If the sun were to undergo this expansion, the Planet Earth would be burned to a crisp. During this period, the pressure at the core, which now consists of compacted helium, "increases to more than 2 billion metric tons per square inch; the temperature rises to 15,000,000°C." As there is no more hydrogen left to ignite, the contraction process continues and the temperature continues to rise. When the heat has reached 100,000,000°C, the helium ignites and a new atomic furnace goes into action. It provides again the outward pressure to counter the inward pressure of the gravitational force. In the burning of the helium, the element carbon is formed. Also, the element beryllium, "which rapidly decays, producing oxygen."

"Eventually", Ditfurth states "the helium nucleus of the star is consumed. Then the whole process repeats itself. The star contracts until it creates even higher temperatures than ever before, at which point the carbon atoms begin to be transformed into heavier elements. The carbon becomes neon and sodium. A complex chain reaction produces new helium nuclei which serve as 'building blocks' for other elements. The chain reaction eventually creates magnesium, aluminum, sulphur and calcium. . . . By this time the temperature of the core has risen to some 500,000,000°C. At this point the atomic process taking place in the core becomes so complex that we cannot adequately describe them here. The atomic reactions produce so much energy that the various elements are constantly being torn down and built up again. Eventually the star exhausts all its store of nuclear energy . . . It has become what astronomers call a 'White Dwarf.' Its extraordinary heat makes it look white."

Ditfurth goes on to tell us that during this long process the star has lost most of its volume, having shrunk to possibly about the size of the Planet Jupiter. What happens next depends on the size of the original star. If it has less than 1.4 times the Mass of our Sun it will remain in its White Dwarf state until it finally becomes a Black Dwarf. It will continue to contract and to cool off as it can no longer trigger atomic reactions. During this long period, however, it has developed, as Ditfurth states, "various heavier elements from the hydrogen of which it was originally composed. But this star never progresses beyond the production of nickel. That is, it produces only about one quarter of the total number of elements. *More importantly, even these elements it produces remain trapped inside the star . . . they stay buried forever in the core of the White Dwarf.* (p. 276, emphasis mine). In time, the White Dwarf becomes a Black Dwarf, just a cinder in the outer space of the Galaxy.

The Ditfurth story of element-formation does not end here, however. The scientists have calculated that stars which have a mass 1.4 times that of our Sun, and larger, will develop a White Dwarf which will have a different ending. Such a star, after having come to the White Dwarf stage, will in the contraction process, develop a gravitational force so powerful that the matter composing the star breaks down. This means that the electron shells in its atoms break down, leading to a "gravitational collapse" of the atomic nuclei themselves. "Before this happens the star is still as large as a good-sized planet. Then, suddenly, in a split second the entire star crumbles away. In a moment it has a diameter of only six to twelve miles! This "implosion" generates temperatures of more than 3,000,000,000,000°C." (p. 277).

At this point we remind the reader that early on we stated that we take the scientist at his word! When this almost unbelievable temperature is reached—200 times hotter than the core of the Sun, an explosion "tears away *one-tenth* of the star's total mass, hurling this matter into space at speeds of up to 6,000 miles per second." We are told that when this explosion takes place, a new star, called a "supernova" is formed. The mass of the new star will be about that of our Sun but it will have a diameter of between six and twelve miles. For a few weeks it will shine with a light 200,000,000 times that of the Sun. The matter composing the star will be densely packed neutrons; a single cubic centimeter will weigh several million tons! The gravitational force of this body, now called a neutron star, may be so powerful that not even photons, light waves, can escape it. The neutron star burns with a temperature of several billion degrees.

The final phase in the life of this star comes, when as it develops new and very heavy elementary particles, the contraction process begins again. Now it will not stop. "The neutron star shrinks away to a purely mathematical point; that is, it becomes a sheer abstraction." Needless to say, scientists do not claim to understand how this star withdraws itself from the Universe and "passes into a realm which is no longer accessible to mathematical calculation." It is now called a "black hole." For our purposes, we will let the story of the life of a star end at this point, even though it does not. One thought should be added. If indeed the mass of such a star "withdraws itself from the Universe"—does not this constitute a breaking of the Law of the Conservation of Energy?

Possibly new energy is constantly being put into the Universe as the "sustained, expressed will of the Creator-God" continues to will that the Universe be! See the discussion in Chapter Four.

Now we come to a point of great interest to our study. Ditfurth states on page 280, "In the course of the gravitational collapse, the star ejected a large percentage of the elements it had created in its core." He actually said that only 10% of the star's mass was hurled into space. He then goes on to state that these elements are caught up in the clouds of interstellar dust and that new stars are born with these elements as a part of their matter. We have problems with that statement. The entire concept of the life cycle of the star is based upon the nuclear furnace in the core of the star. The first fuel was hydrogen and the star produced only helium. In the next phase, helium became the fuel for the nuclear furnace and carbon and oxygen were produced. When the helium was exhausted, carbon became the fuel. We have been told that this process continued to repeat itself until all nuclear fuel was spent. We must ask: if each element that was involved as fuel in the atomic fires was burned up in the formation of heat and pressures which were involved in the formation of new elements, how many elements were actually present in the core of an exploding star? Certainly many elements must have been missing. Also as we have just indicated, we are told that only 10% of the mass of a supernova was "hurled out into space." What does that say with respect to the amount of elements, as well as number, that are put into space when a supernova explodes? Add to this uncertainty the realization that these explosions of supernova occur only occasionally and that they must be at points many trillions of miles from each other. The concentration of elements in outer space cannot be heavy. It would be quite unlikely that one potential solar system cloud would randomly pick up all of the 103 natural elements before developing what we call our Solar System, composed of a sun and nine planets!

Yet, Ditfurth claims, "we now know that many successive star generations have come and gone, each of them forming from the matter provided by the previous generation. *All the matter existing in the various spiral galaxies has been used over and over;* it alternately condenses into stars and is released again into space. Gradually the stars have manufactured all the heavy elements in the periodic table of the elements, including the heaviest of all, uranium." The statement sounds like pure speculation. We have been told that 90% of all the stars in our Galaxy are about the size of our Sun or smaller. This means that no elements that might have been formed in these stars will ever get out into space—they will be locked up forever in the Black Dwarf stars. Their matter will never be recycled through "successive star generations." The remainder of the stars, while larger in size, represent only 25% of the total mass of the Universe, according to Gribbin, and these stars, exploding as nova or supernova stars only discharge from one to ten% of their mass into space. The amount of matter available for recycling becomes relatively small.

There is another difficult problem to struggle with in respect to the concept that the elements of the Universe have all been formed in nuclear fires in the cores of the billions of stars as they have been recycled from one generation to another. We are given the account that the life story of all stars is basically identical—they form from huge concentrations of hydrogen gas that separate out as distinct clouds and begin to contract in on themselves. The core that is developed in the center of each cloud becomes so hot that the hydrogen nuclei begin to fuse and the star is born. In time the supply of hydrogen is exhausted. The atomic fire goes out. Gravitational force exerts its pressure and the core begins to contract again. This causes great heat to develop. When the temperature has reached about 100,000,000°C the helium which has been formed from the hydrogen becomes the nuclear fuel. When the helium is gone, carbon, newly created in the helium-burning period, becomes the fuel. This process continues until all nuclear fuel is exhausted. Then, if the star is about the size of our Sun or smaller, it goes through a lengthy process which ends with it becoming a Black Dwarf, a dead star in space. If the star is larger, say eight or more times the mass of the Sun, the process leads to the development of a supernova which ejects newly created elements into space as it explodes—as we have stated earlier. The mass of the core of the supernova, however, is retained while the volume shrinks. It becomes a neutron star. The contraction process under the absolute power of the gravitational force continues and the star, as Ditfurth puts it, "simply exits from the stage. In some incomprehensible manner, it literally disappears from the universe," becoming a black hole.

If this all be so, how can we account for the globular clusters, the oldest stars of the Universe? We are told by Ditfurth that early in the history of our Galaxy, before it had taken its present shape, these stars formed out of the hydrogen gas of the Creation Cloud as it broke up into what we call the galactic clouds. "We can still identify the most ancient suns of our stellar system. They developed when our Milky Way Galaxy had not yet developed the shape it has today; that is, they were born when our galaxy still had a roughly spherical shape." The oldest stars form the "so-called 'globular clusters.'" Thus far astronomers have discovered 119 such globular clusters in our galaxy. They are evenly distributed on all sides of the center of the galaxy, composing a sphere. Astronomers believe that these ancient stars mark the space occupied by our Milky Way Galaxy when it was still a spherical ball of gas... Their age has been estimated at somewhere between 6 and 10 billion years. They appear to consist of pure hydrogen and helium, *with no admixture* of heavier elements" (p. 284). John Gribbin strengthens this statement when in a discussion of globular clusters he says, "Our best theoretical models of stellar evolution suggest that the stars in globular clusters, the oldest known stars, are just about this old (10 billion years) and perhaps even older" (*Genesis*, p. 331).

What do these statements do to the generally accepted life story of a star? Our hydrogen burning Sun is looked upon as an average sized star. It is about 4.5 billion years old and it is believed that it has enough nuclear fuel to last another 4.5 billion years. If it were larger than it is, the hydrogen would have to be burned at a faster rate to offset the additional gravitational force that would be exerted on a star of greater mass. This would give it a shorter life span. This is an important point made by the astronomers in defending the theory of "successive star generations" being responsible for the creation of new elements. The larger the star, the shorter the life span will be. Thus, the globular clusters cannot be made up of stars larger than our Sun. But, neither can they be much smaller, for they have a luminosity that would require a greater mass in order to be observed as they are. This presents a mystery. They should have run through the life cycle of a star long ago. As these stars are still observable, their hydrogen nuclear furnaces must still be going strong. This would indicate that they may yet have as many years to live as they have already enjoyed! The significance of this is that this would give them an unusually long life span—possibly 20 billion years and up—which means that these stars are not operating in accord with the star-life-pattern that we have been given.

There are likely many billions of stars in the globular clusters in our Galaxy as more clusters are being found as time passes. Some of the clusters contain million of stars. Our nearest galactic neighbor, the Andromeda galaxy two million light years distant, has more than 200 clusters so far identified. Other galaxies also show the presence of globular clusters. This means that very likely, after the break-up of the Creation Cloud into the galactic clouds, these were the first stars to develop—and they are still functioning! Thus, we are forced to ask the questions. Should not these stars have long ago lived out their life-span and become Black Dwarf stars? Can it be that the whole concept of the life history of a star, as it has been given to us, is inaccurate? Could it thus be that the whole story about element formation by stars is not adequate because we don't have the whole story about stars? To this observer, the answers to these difficult questions can only be found through accepting the concept that from time to time the Creator-God intervened to expedite the fulfillment of his purpose in developing a "home" for life on the Planet Earth. Earth is the only celestial body known that has all 103 elements. Scattered in the vastness of intergalactical space, many of the elements have been identified—but not all of them. How they all came to be concentrated on one tiny planet inside a "closed" Solar System, is indeed a mystery!

One thing we know for sure. We live on the Planet Earth. We enjoy life. We have light and warmth. We have food to eat, water to drink. We have an abundant plant life, fish in the waters, birds in the air, animals on the land. As we shall see more clearly a bit later, plant life is essential to make possible animal life and animal life is an absolute necessity for continued plant life. The life systems are supportive of each other; neither could exist without the other. We have seasons in our year which contribute to the well-being of all forms of life. We have energy stored in huge quantities to supplement the energy of our Sun. Life on our Planet is good, for the Planet was supplied with all the substances and conditions necessary to develop and sustain life.

The scientific evolutionist declares that the Universe and all that is in it came into being by happenstance, by the random movements of atoms in the primordial Creation-Cloud that came into being as a result of the Big Bang. They do not explain how the Big Bang, with its enormous supply of energy came into being. They want us to believe that everything has happened in accord with the Laws of Nature, but they deny that a Law Giver placed those laws in the Universe to serve it. They would have us believe that this Planet just happened to form in what we call the Solar System, a tiny speck in the Universe, too small to be noticed by anyone who might be somewhere else! This Earth also just happened to have all 103 of the natural elements. The scientists have not provided adequate evidence to explain the existence of these elements—to say nothing about how they all are to be found on Earth. They have warned us, however, that to believe that any Intelligence, or Supreme Being, had any part in this marvelous development or that such a Power set in place the Laws of Nature which govern all the Universe, is simple anthropomorphic thinking and that such thoughts should be put out of mind!

Those who believe that a Creator-God, who has always existed and who will exist eternally, brought the Universe as we know it into being by the exercise of his will, believe also that with all the rest of the Universe, the Planet Earth was a part of his grand design. The Creator-God set in place the Laws of Nature which under his will—and with some "direct assists" from time to time—brought into being all of the elements necessary for the development of the Universe, including our Galaxy, our Solar System, and our Planet Earth. While the mystery of the formation of the 103 elements of the Universe has not been resolved, we do have a better understanding of the problems involved. We also have a plausible explanation of how it all happened.

CHAPTER VIII
THE PLANET EARTH, PART II: THE MYSTERY OF THE PROTECTIVE SYSTEMS

We have another great mystery to probe as we look at the Planet Earth: the mystery of the protective systems which shield the Planet from deadly cosmic radiation and so help to provide on Earth a home for the life that was to be developed. As we begin this discussion we should seek to get a picture of the Solar System in our mind. It is like a huge sphere with a diameter of about ten billion miles moving in a stately fashion in a great orbit around the center of the gravitational force of the Milky Way Galaxy. The Sun is at the center of the sphere and thus is its gravitational center. Nine planets are in orbit around the Sun, the Planet Earth being the third out from the Sun. When we speak of the Sun being at the center of the Solar System sphere we do not mean the "absolute" center, for the sphere has a kind of life of its own; it is not exactly the same in shape at all times, but is generally so. The sphere is in orbit, moving through the interstellar space of the Milky Way Galaxy at the astounding speed of about 162 miles per second! It would be difficult to grasp the hugeness of the volume of power that must be involved in the movement of so vast a body with so much mass at such a high speed. The orbit around the center of the Galaxy takes about 250,000,000 years. In its lifetime, the Solar System has made only about 20 trips around the center of the Galaxy. If it could be seen from some point outside of itself, our Solar System would present a grand and awesome picture. As it moves through space, however, it would be in constant danger were it not for an amazing protective device that shields it from the deadly cosmic radiation which fills all of the space of our Galaxy. We will try to explain how this device works. Besides the huge amounts of non-material electromagnetic radiant energy that streams from the surface of the Sun into every part of the Solar System, the Sun also emits a second form of radiation which is corpuscular in nature. It consists of a stream of energy particles—protons and electrons. "When these particles leave the sun's surface, they are traveling at a speed of more than 310 miles per second. As they fly past the earth several days later, they are still moving at almost a thousand times the speed of sound" (Ditfurth, *Children*, p. 72). This stream of particles is known as the "solar wind." For many years scientists had been seeking to find answers to some mysterious observations. Why do comets appear to be pushing their tails in front of them as they move away from the Sun? What causes the "northern lights" to shine at both the northern and southern magnetic poles? "As early as 1896 the Norwegian physicist, Olaf Birkeland, had theorized that the northern lights were produced by 'corpuscular radiation' or some sort of 'wind' from the Sun. This 'wind,' he thought, must consist of tiny electrically charged particles" (Ditfurth, p. 76). Other scientists more recently were coming to believe that the peculiar action with respect to the comet tails was caused by small electrically charged particles being hurled out from the surface of the Sun, so that the tail of the comet always had to be pointed "away" from the sun, even when the comet was leaving the region of the Sun. It appeared that something like a "wind" controlled the position of the tail of the comet.

The breakthrough on this mystery came when lunar probes were launched in the space program. The Russian and American probes both verified the existence of the "solar wind." The lunar probes discovered a high radiation zone which formed a belt around the earth's equatorial region beginning at a distance of about 600 miles out from the earth's surface. The radiation increased in intensity until about 3,000 miles out, after which it decreased rapidly. Again, at an altitude of about 12,500 miles, a second radiation belt, much larger than the first, was discovered. It stretched all around the earth except for an opening at each pole. Our diagram will illustrate. These zones are now called the Van Allen radiation belts in honor of the man who was responsible for finding them. "Investigation revealed that both belts were composed of concentrated electrical particles. The wide upper belt consisted primarily of electrons, the lower belt of protons. But where did these energetic atomic particles come from? Their only possible source was the Sun. It remained to be discovered how the particles found their way into the upper layers of our atmosphere" (Ditfurth, p. 80).

Ditfurth goes on to explain the origin of the Van Allen radiation belts. "Both belts are composed of protons and electrons which originally traveled to the earth in the form of solar plasma (solar wind). How could these particles have slipped through our magnetic defense system? . . . Probably the breakthrough of solar plasma takes place through the rear, or 'tail' of the magnetosphere. So much turbulance exists in this area that it no longer maintains a solid barrier against the solar wind. Thus a few particles literally infiltrated our defense system from the rear" (p. 102). But if this is so, and very likely it is, how do they become established in two clearly defined belts—the positively charged protons closest to Earth and the negatively charged electrons at a distance about 10,000 miles farther out? It should be noted that

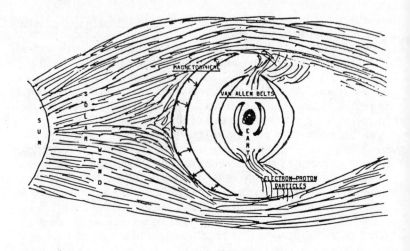

THE MAGNETOSPHERE

The earth is surrounded by lines of force set up, scientists believe, by the fact that the metallic core of the earth acts as a dynamo creating an electrical current due to the motion differential between the core and the crust of the earth. The lines of force constitute a huge magnetic field. Ditfurth, on page 97 in *Children of the Universe*, states, "The lines of force arise almost perpendicularly from the poles, gradually curve along a horizontal line, and finally descend in an almost perpendicular line to the opposite pole." This magnetic field is called the *magnetosphere*. It reaches out about 55,000 miles from the earth on its outer side while the inner side is about 22,000 miles out. It protects the earth from the deadly Solar Wind. The Solar Wind, composed of negative electrons and positive protons, cannot penetrate a magnetic field. Thus when the Solar Wind strikes the magnetosphere it is shunted around the earth—as shown in the diagram illustrating the action of the Solar Wind. The Lunar Probes found that the zone of the magnetosphere was almost free of Solar Wind particles or plasma.

there is a rather mysterious correlation with the structure of the hydrogen atom—the electrons orbiting about the inner core of protons. It would appear that the assembly by the Van Allen belts as well as their functions, constitutes another bit of evidence of design with a purpose. The magnetosphere defense system involved in this matter will be discussed a bit later in this chapter.

"The sun emits huge quantities of electromagnetic radiation, above all, light and heat. But it also emits corpuscular radiation in the form of protons and electrons which fly out from the sun at more than a thousand times the speed off sound. Satelite data indicate that each second the sun pours no less than 1,000,000 metric tons of matter into the solar wind.... The expression solar 'wind' is highly appropriate. It makes clear that we are dealing not with ordinary radiation, but with the emission of actual physical particles"—the wind off protons and electrons blowing off the sun. (pp. 80–81).

The classification of the pre-atomic substance has been a problem. For the electrons and protons are not true matter until they come into the proper relationship with each other. Considering electrons and protons as "particles" implies that they constitute units of matter. This cannot be strictly so, for they are "energy" forms and matter as such does not appear until an electron and a proton are joined by way of establishing the atom of the element hydrogen—one electron in a sustained orbit around one proton. Apart from such a relationship between electrons and protons there are no elements, there is no matter. However, it is not as simple as that, for the energy forms, electrons and protons—and we could also add neutrons—act like they are particles of matter. They move, they have mass, they carry an electrical charge (except for the neutron). As we will describe in Chapter Ten, the mass of the electron represents a measure of the attraction between the electron and the center of gravitational force as it is exercised by the Will of the Creator-God. Thus we give the electron a number 1 ugf representing the power-attraction level between an electron and the gravitational force—one unit-of-gravitational-force. The proton has a "mass" 1836 times that of an electron and therefore we give it a ugf of 1836. Thus both electrons and protons act as though they were a "measurable something" and thus they can be thought of as packets of energy or "particles". The proton, as we have learned, can be divided. It is in this sense that Ditfurth can speak of "corpuscular radiation in the form of protons and electrons." Without such an understanding, the term "corpuscular radiation" would be meaningless. Thus, we can use the term "particle" in reference to electrons and protons, even though they are not true matter as we know it.

As the solar wind moves past the Earth it is traveling at a speed of 186 miles per second. It sweeps the Solar System relatively clear of interstellar matter consisting of hydrogen gas and dust. This action applies a kind of brake on the speed of the wind. As it continues on its course, it not only slows down but becomes thinner as it must be expanding continuously in all directions in order to fill the huge volume of space in the solar system sphere. The protons in the wind carry their positive electrical charge and the electrons carry their negative charge. By the time the solar wind blows beyond the orbit of Pluto its strength is greatly diminished. The wind has reached the edge of the Solar System. At this point it comes into physical contact with enormous amounts of radioactive interstellar dust and is no longer strong enough to sweep it out of its path. When this happens, it causes magnetic storms to break out at all points around the outer edge of the Solar System sphere. These storms are created by the interaction of the electrically charged particles of the solar wind with the highly radioactive cosmic dust. Thus, all around the Solar System sphere there are electromagnetic storms raging constantly. These storms develop powerful "lines of force" which in turn form a kind of buffer zone between the Solar System and the true outer space of the Galaxy. "Such a shock or border zone would form a huge sphere surrounding our entire solar system. The magnetic storms caused by the clash of solar plasma with the interstellar dust would arise on all sides at an approximately equal distance from the sun" (Ditfurth, p. 89). The turbulent area would be from several hundred to possibly a few thousand miles thick. "Thus our entire solar system may be inclosed within a huge sphere. This sphere would not only be invisible, but also incorporeal. The solar particles are not nonmaterial, but the magnetic storms they create are. We may all be living under the protection of an invisible and nonmaterial force" (p. 90). We recognize that Ditfurth is using the word "may"—but as in all subsequent reference to the protective system created by this nonmaterial force, he accepts it as being factual, we also accept it as fact.

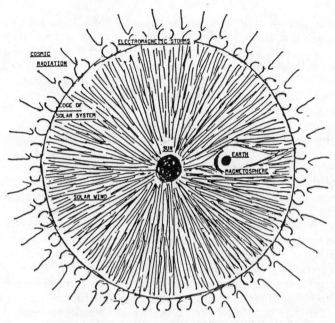

THE ELECTROMAGNETIC BUFFER ZONE

The Solar Wind, blowing at the rate of 310 miles per second, leaves the Sun on its way to the very edge of the Solar System Sphere. The powerful wind is composed of negatively charged electrons and positively charged protons. A million tons of energy particles leaves the sun each second! As it moves out into the space of the Solar System the corpuscular wind sweeps the space relatively clear of dust and other debris. As it reaches the outermost edge of the Solar System Sphere it engages clouds of interstellar dust, highly radioactive. The collision results in huge electromagnetic storms, surrounding the entire Solar System. This has the effect of creating an electric field of force, a magnetic field, which ranges from several hundred to several thousand miles in thickness all around the Solar System Sphere. The diagram is just two dimensional—we must visualize it in three dimensions. Cosmic radiation coming from outer space cannot penetrate the magnetic field, it is turned back into space, being reflected like a ray of light is reflected by a shining surface. Thus, the entire Solar System Sphere is protected from deadly radiation which otherwise would have devastating effects on the Planets of the Solar System. The diagram also shows the location of the magnetosphere—a second protective device shielding the earth from the Solar Wind.

Outer space—that which is beyond our Solar System—is filled with cosmic radiation which represents the most intense radiation ever measured. The particles composing it travel at almost the speed of light—no more intense radiation could possibly exist. The significance of the Solar Wind becomes clear at this point. The buffer zone that surrounds our Solar System sphere protects it from the deadly radiation that would otherwise bombard the Earth and all else within the Solar System. It does this by reflecting most of the cosmic radiation-particles, like a mirror reflects light rays, turning them back into outer space. As these discoveries have been made it has been realized that the rise and fall of the amount of cosmic radiation that does reach our Planet can also be explained by the fact that solar flares or eruptions on the surface of the Sun increases the number and the speed of the particles of the Solar wind. This in turn causes the effectiveness of the border zone magnetic field to fluctuate. Were it not for the Solar Wind streaming from the Sun to the outermost edge of the Solar System, where it produces magnetic turbulance in its clash with interstellar cosmic dust, creating vast magnetic fields around the entire Solar System and thus shutting out the cosmic rays, the Earth and all the other Planets would be burned. No life would be possible on Earth.

Thus, the beneficial effect of the solar wind is very great. It sweeps the interior of the Solar System sphere relatively clear of interstellar matter as it rushes out to the edges of the Sphere; then it performs its greatest task when it creates the nonmaterial magnetic field force that shunts away the cosmic rays, turning them back into outer space, saving the Planets of the Solar System from that most violent radiation. We ask the question again. Can it be said that this huge defense system protecting the entire Solar System from deadly radiation has come into existence by mere happenstance?

But, what about the solar wind itself? It also is a violent force as it hurtles out from the Sun at 310 miles per second. We find that there is a second "line-of-defense" which protects the Planet Earth from the Solar Wind! This defense is the magneto-sphere referred to earlier when we were discussing the Van Allen radiation belts.

The Earth is surrounded by a huge magnetic field, caused, scientists believe, by the fact that the metallic core of the Earth acts like a dynamo in the creation of an electric current as the Moon causes the ebb and flow of the tides—which in turn, causes a movement of the crust of the earth relative to the metallic core. This is an oversimplification of the matter, but it must suffice at this point. The lines of force which always arise from an electric current compose the magnetic field that surrounds the earth, originating at the two poles. "The lines of force arise almost perpendicularly from the poles, gradually curve along a horizontal line, and finally descend in an almost perpendicular line to the opposite pole" (Ditfurth, p. 97).

Because the Solar Wind particles are electrically active, they are susceptible to magnetic influences. Thus, as Ditfurth continues, "The earth's magnetic field 'controls' the solar wind by simply shutting it out. Satellite investigation showed that the magnetic field (surrounding the Planet) contains almost no particles of solar plasma." This zone is called the magnetosphere. "Thus we live under the protection of two invisible spheres, one fitting inside the other. The first sphere is generated by the solar wind: it consists of the shock zone extending beyond the orbit of Pluto, that owes its existence to the clash between solar plasma and interstellar matter. It envelopes our entire solar system. The second sphere is much smaller but equally important to human beings. It consists of the magnetosphere, which shields the earth from solar plasma" (p. 99). We would add a third protective device, the Van Allen radiation belts, inside the magnetosphere, which screens out cosmic rays which have escaped the other systems.

Inside the magnetosphere are the Van Allen radiation belts. Ditfurth suggests that they have been formed by a leak in the magnetosphere defense system at the far side of the Planet—and this is entirely possible, as our diagram illustrates. The mystery, however, is that these particles of the solar wind, electrons and protons, arrange themselves neatly in two orderly belts which are located in the earth's upper atmosphere. The protons are on the "inside" at about 600 to 3000 miles out from the earth and the electrons are on the "outside" at a distance of about 12,000 miles. How they arrange themselves in this manner and how they maintain this kind of structure is not really explained by our scientists. The belts are highly radioactive and do a very important job. They capture most of the cosmic rays that have managed to escape the buffer zone which encloses the entire Solar System Sphere. It

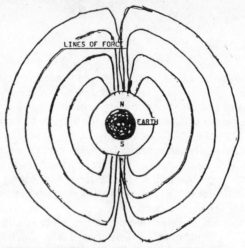

THE EARTH'S PROTECTIVE SYSTEMS

The Solar Wind leaves the Sun at a speed of 310 miles per second. If it should reach the Planet Earth it would destroy all possibility of life. This does not happen, however, because of a protective device built into the atmosphere of the Planet. The magnetosphere provides an effective shield for the Earth. The lines of force, constituting a magnetic field cannot be penetrated by the electron-proton particles of the Solar Wind. The magnetosphere acts like a rubber cushion which is depressed by the force of the wind—which travels at a constant high speed but with varying strength—the variations caused by the sun-spot activity on the surface of the Sun. The magnetosphere thus may vary in thickness from 15 to 20 thousand miles as the lines-of-force are closer together. The Solar Wind is shunted around the Planet by the presence of the magnetosphere—as shown in the diagram. As the Wind passes the earth it pulls with it the lines-of-force which are on the far side—away from the Sun—and carries them out for possibly 600,000 miles. This "thinning out" of the magnetosphere allows some of the solar wind plasma—electrons-protons—to come back into the atmosphere of the Earth. In a manner that is not really understood, the electrons form a belt about 12,000 miles out from the surface of the Earth and the protons form a much smaller belt about 4,000 miles out—as the diagram shows. These are the Van Allen radiation belts which constitute a third protective device for the Planet as they serve to trap radiation particles which have escaped the first two devices. They trap this radiation and allow it to escape only at the two poles where it shoots back into space, away from the Earth! These shield systems are entirely invisible, we don't see them, but they are real! How these two radiation belts assemble as they do, in two well-defined belts and how they are maintained, are questions that as yet have no clear answers—except that it fits the design and purpose of the Creator-God.

appears that they "bottle up" the cosmic rays which cannot penetrate the electrical field set up by the charged electrons and protons. The cosmic rays retain their immense energy and move from pole to pole, seeking an escape which does happen occasionally. However, when cosmic rays escape this trap it can happen only at one of the poles and thus they shoot back into space and away from the Planet Earth! Were it not for the Van Allen radiation belts a statistically significant number of cosmic rays would strike the Planet, causing severe damage.

Thus, there are three protective devices that shield the Planet from deadly cosmic radiation. Together, they constitute a protective system without which life would not be possible on Earth. To hold that these devices are the result of random happenings without design or purpose appears to this writer to be an absurdity. Summarizing them, we say:

THE SOLAR WIND, blowing from the surface of the Sun, causes electromagnetic storms at the point where it engages cosmic rays at the outside edge of the Solar System Sphere—about five billion miles out from the Sun. These storms set up a thick electromagnetic field, a buffer zone around the Solar System, which reflects cosmic rays back into the outer space of the Galaxy.

THE MAGNETOSPHERE, which surrounds the Planet Earth at a distance about 50,000 miles out, is formed by the magnetic lines of force set up by the dynamo action of the metallic core of the Earth moving in relationship to the crust of the Earth. The tidal effects of the Moon as it exerts its gravitational pull on the oceans of the earth, set up the motion differential between the core of the Earth and the crust. This creates the electrical current which produces the lines of force. The Solar Wind, consisting of electrically charged particles, cannot penetrate the lines of force and is thus effectively shunted around the Planet.

THE VAN ALLEN RADIATION BELTS, located within the magnetosphere, screen out most of the cosmic rays which have penetrated the outer buffer zone. The cosmic rays which do escape at both the northern and southern poles produce the Aurora Borealis, the lights of many colors that shimmer in the northern and southern skies.

These three protective devices constitute a major "system" which has as its purpose the protection of the Planet Earth from deadly radiation which otherwise would preclude the possibility of life on Earth. There are other protections built into the scenario of life on Earth such as the very bulk of the Sun's mass which diminishes greatly the strength of the radiation coming out from the core. The very atmosphere of the Planet which cuts down on the strength of the ultraviolet rays and other energy rays which would prove detrimental to life, is another important protection of life on Earth. No other Planet enjoys these protective systems. No other Planet supports life. Were these protections set up by happenstance, by the "willy-nilly" movements of atoms or their component parts? Such a position as is taken by the scientists does not appear to be a possibility.

CHAPTER IX
THE PLANET EARTH, PART III:
THE FORMATION OF THE PLANET

The scientific writers appear to go out of their way to impress upon readers that the Planet Earth is no big deal. It is certainly not the center of the Universe, as people a few hundred years ago believed. In fact, they tell us our Sun is just an average sun, there are billions more just like it. Our Solar System is actually so small that in terms of Universe space, it would be almost impossible for an astronomer in a different solar system out in space to locate it. Even our Milky Way Galaxy must be downgraded to the status of average, among the billions of galaxies which make up the Universe.

While all of this may well be true, there are some things to be said on the other side. Being average may not be all bad—for as we have seen, the larger stars have short life spans; the smaller stars cannot develop life support systems! Our Solar System is a huge sphere, containing our Sun which is 864,000 miles in diameter around which nine planets are in orbit. The Solar System Sphere—about ten billion miles in diameter—is protected from deadly cosmic radiation by an electromagnetic band ranging in thickness from several hundred to several thousand miles. Within the Solar System Sphere there is a Solar Wind streaming from the surface of the Sun which serves to sweep out of the Solar System interstellar dust and debris which could be harmful. This it does while on its way to the outer edge of the Sphere where it is responsible for setting up the electromagnetic protective band. The Solar Wind in itself would be a deadly force if it were to strike the Planet Earth. No life would be possible on earth if this wind were to blow across its surface. Around the Planet, however, at a distance ranging from twenty to fifty thousand miles out, there is a huge magnetosphere which is really the earth's magnetic field set up by the electrical current which is developed by the pull of the Moon on the crust of the earth, as described earlier. This magnetic field shuts off the Solar Wind, shunting it around the Planet. No other planet in the Solar System has this protection—and none of them support life. Below the magnetosphere—or within it—are the Van Allen radiation belts which trap stray cosmic rays that have escaped the other protective devices.

All of this would indicate that the Planet Earth is not just an ordinary body in space. It evidently has a purpose and is part of a design related to the development of and sustaining of life. There are a number of other factors which separate Planet Earth from the other planets of the Solar System. It has an atmosphere made up of nitrogen (78%), oxygen (20%) and other, rarer gases. It has an ample supply of water; its temperature is moderate, averaging about 15°C. Thus water won't boil away as it did on Mercury and Venus nor will it freeze solid as it does on Mars. It has a rich soil in which plants can grow. Gribbin, in comparing Earth with Venus and Mars, states in *Genesis*; "The Earth alone of the three has an atmospheric blanket which is just right to keep the surface of our planet in the temperature range between the melting point of ice and the boiling point of water" (p. 115). The Planet possesses all of the 103 natural elements, in quantities, locations and forms that make them usable by plants and animals. It is, in short, an ideal place for the purpose of developing and sustaining of life as we know it. The scientific evolutionist insists that all of these remarkable coincidences came about quite by accident, with no help provided by a Creator-God.

"The most important thing we know about the origin of the Earth is that our home planet formed at the same time as the Sun, with the rest of the Solar System, from a collapsing cloud of gas in interstellar space" (Gribbin, p. 107). We agree that this is an important fact. It tells us that the Earth was developed out of the same stuff as the Sun and the other planets, a cloud of gas that was 75% hydrogen and 25% helium, plus a tiny smattering of other elements. This means that the content of the cloud was very much like that of the original Creation-Cloud that resulted from the Big Bang. To be sure, it had been a part of the Milky Way Galaxy for a long time before it separated out as a cloud that was destined to become a Solar System, with a Sun and nine Planets. During that time it had no doubt picked up some of the trace amounts of elements that were in the Galaxy Cloud, debris from exploding Supernova. This is in line with the statements of the various scientific writers already quoted. Gribbin, however, continues with a startling statement: "The cloud of dust and gas *within the Solar System which collapsed to form the Planet Earth was probably composed chiefly of silicon compounds, iron oxides, and magnesium oxides, with just a trace of all the other elements we now find in our terrestrial environment*" (p. 108, emphasis mine).

Gribbin implies in this statement that the Planet Earth was formed from a cloud of dust and gas which in some way was set apart "within the Solar System". He says that this cloud, upon collapsing, formed the Planet. He also said that this cloud had in it all the elements found on Earth. No other writer that I have read would go along with that statement. Following through on the concept, does Gribbin also imply that each of the Planets in the Solar System was formed separately by the collapse of their own distinct clouds? This concept would give us a very different picture of the Solar System Cloud that moved out from a spiral arm of the Milky Way Galaxy than that of an homogenous cloud composed of two gases, hydrogen and helium.

Possibly, Gribbin is trying to explain how the Planet Earth has the composition that it has. If this is so, he must explain an even greater mystery—how one cloud of "dust and gas" got into the Solar System Sphere, "composed chiefly of silicon compounds, iron oxides, and magnesium oxides, with just a trace of all the other elements we now find in our terrestrial environment." It would have been easier to believe that the entire Solar System cloud had such a composition than that a cloud that had originally been homogenous would have through the passage of time differentiated into possibly nine clouds with unique compositions.

Whatever significance may be attached to Gribbin's account of a separate cloud of dust and gas collapsing to form Planet Earth, we find that he reverts to the quite widely accepted view as to the actual formation of the Planet. It happened because particles within the Solar System Cloud collided and stuck together. As the resulting "larger" particle gathered mass, it grew by accretion at an ever more rapid rate because its gravitational pull attracted smaller particles and held them to the growing proto-planet. Here we are reminded of the problem already discussed, as to how it could happen that in the large volume of the Solar System Cloud, only nine such Planets were formed—only one of which is suitable for life.

We recognize that it is a difficult task, about four and a half billion years after the fact, to describe exactly how it all happened. We do believe, however, that the scientific evolutionist must be held to his ground. The premise of the scientific community is that all of the Universe in its many parts, including our Solar System and our Planet, has come into being by the random movements of energy particles in the Big Bang cloud, followed by the equally random movements and collisions of atoms in the subsequent bodies in space. The scientist might seek to counter this statement by pointing out that the assembly of the material that became the Planet Earth was not a random action but one that followed the laws of nature. In doing that, however, he must also explain how the Laws of Nature came into being out of the chaos following the Big Bang explosion. This he never does. It is as though he feels that he has a right to accept these laws as a given! If he would actually recognize this, we might not have a problem—for if the Laws of Nature are a given, the existence of the Giver must be acknowledged.

Our position is that the Sun and planets developed as they did because the elements of the Universe were under the control of the Creator-God who in the exercise of his will set in place the Laws of Nature, reserving the right of intervention from time to time in directing the course of events. A very likely point of intervention in this instance is that while the Solar System Cloud was 99.9% taken up by the Sun, the balance, 0.1% was separated out in such a manner that the Planet Earth received all of the

103 elements in amounts sufficient to meet the needs of that body in the development and maintaince of life. This would also call for another earlier intervention in which the 103 elements were provided to the Solar System cloud in whatever manner was in the will of the Creator-God. As we have seen in an earlier quote from Davies, the concentration of these heavier elements on the Planet Earth is highly atypical of their abundance in the Universe at large. We are stating strongly that it was not by happenstance that in the vastness of space our relatively tiny Solar System Cloud found itself with all the ingredients necessary for the development of our present Sun, with nine planets, one of which supports life as we know it.

Gribbin tells of the development of heat in the new planet. As the particles of gas and dust collided, the first heat came from the friction that built up in the interior of the proto-planet. "Then, as a planet proper began to grow by accretion, the interior was pressed ever tighter by the increasing weight of the material above. Just as a gas cloud warms up as it contracts, releasing gravitational potential energy as heat, so did the proto-earth" (p. 108). The planet was being "created" by the steady accretion of gas and dust—with some of the particles having developed into quite large chunks—and the picture we get from this is that everything was in a state of flux. The body of the planet was molten, a plasma, in which material stuff could separate out in accordance to weight. Heavier matter was sinking toward the center, the core, while lighter material was rising to the surface. The solid crust must have developed later as the planet reached its full size and became more stable with the outer surface cooling. The second source of heat was the decay of the radioactive elements like uranium and thorium, which must have been brought into the Solar System cloud by the explosions of supernova star at some time prior to its contraction. As these elements decayed—that is, broke down into simpler elements which were more stable—they produced huge amounts of heat by the process of nuclear fission. It was likely that this source of heat provided the temperature necessary for the melting of iron and nickel, both of which sank toward the center of the proton-planet Earth. As these heavy metal globules moved toward the center of the forming planet they also gave up gravitational heat in the process. Thus, the structure of our Earth as it now is began to take shape.

It is a sphere with a diameter of about 7,920 miles. The very center of it is believed to be a core of solid iron—solid because of the great pressure—and this core is surrounded by a liquid, a molten layer of iron-nickel. Around this liquid core is a layer that is called the dense mantle, made up of materials lighter than the inner core. Above this layer is another, called a transition layer, and above that is the light mantle. Then comes another transition layer. Then comes the final surface or crust, made up chiefly of silicon, aluminum and magnesium compounds, floating on the surface. Very little is known of the chemical composition of the mantles.

The scientific writers do not tell us "how" the chemical reactions which brought into being the contents of the planet occurred. When and how was the iron created; the silicon, magnesium, and aluminum, whose oxides form the outer crust of the planet? It would appear that the Earth's interior, as it formed, must have been a huge cauldron of molten material with chemical activity producing what we now know as the stuff of the Earth. Again, we are thinking of the scientific description of the contents of the Solar System cloud—75% hydrogen, 25% helium, with just a smattering of heavier elements comprising 0.1% of its total mass. It is bizzarre to think of the hugeness of the volume of the Solar System cloud and the relative insignificance of the volume of the planet, and then to accept the scientific position that whatever happened to accomplish the establishment of the planet was purely happenstance, with no guiding force, no intelligent plan or purpose being involved. The 0.1% of heavier elements somehow were concentrated primarily on the four terrestrial planets with only the Earth having all 103 elements! Why the gravitational force of the Sun did not pull all these elements into itself is an unexplained mystery.

The Planet Earth is thus composed principally of solid material—even though much of it is in a molten state. Most of the outer crust is in the form of oxides of silicon, magnesium, iron and aluminum. The inner mantles are composed of gas and other substances but we do not know their chemical make-up. The outer core is molten iron-nickel and the inner core is solid iron. This tells us that there was design and purpose in the development of the Planet, for the iron core provides the basis for the provision of the magnetosphere which surrounds the Planet like a protective shield. It also provides the gravitational force at the center of the planet which in turn is subject to the gravitational force at the center of the Galaxy. In the growth of the Planet out of whatever material was available to the Solar System Cloud, the Laws of Nature as set up by the Creator-God were in charge—everything was done in an orderly manner.

The heavy metals, iron and nickel, settled in the center of the globe, giving stability and strength to the young Planet. The lighter elements, in whatever chemical combinations, were located in their proper positions with respect to their weight. But not entirely so—for tin, copper, zinc, iron, silver, gold, lead, uranium, are found in good quantities up in the crust. Surely they were brought there by volcanic action and the upward thrust caused by the pressure of the molten plasma of the mantles—but even so, they represent aberations in the Laws of Nature—like some force had intervened for a purpose. It was necessary to have these heavier elements—and all the others—on the surface available for use on a Planet which was to develop plant, animal and human life.

"The continents are made of light rock, chiefly granite, while the material of the crust beneath the oceans, chiefly basalt, is rather heavier, so that to the very end the lighter materials were rising higher than the denser materials, in line with the simple laws of physics. We can even put a date on when all this happened, since the oldest rocks on Earth are 3,900 million years old, and the age of the Solar System is about 4,500 million years" (Gribbin, p. 110). So, in rough terms, it took about 600 million years for the proto-Earth to develop a hot, molten interior, for "differentiation" of light and heavy material to take place, and for the crust to begin to solidify—just about 13% of the history of our planet to date." It is interesting to note how consistent the scientist is in his position that no Designer was involved. It all happened "in line with the simple laws of physics"!

Thus, we get a picture of a huge, round body in orbit about the pale yellow new Sun of the Solar System. It had a very thin crust, thinner than the skin of an apple relative to the size of an apple. The surface was made up of the lighter materials, mainly silicon, aluminum and magnesium compounds. At this point in time there were no oceans of water. There was no atmosphere and thus no wind. It must have presented a desolate picture. In the inside, however, it was hot and getting hotter as the radioactive elements broke down, releasing enormous amounts of heat. The turbulance of the molten material, pushing up from the interior, with ever increasing pressure, caused volcanos to erupt with great violence, throwing up huge amounts of material high into the airless space above the surface. Clouds of gas and water vapor began to form the first atmosphere—mostly carbon-dioxide and water vapor to begin with. It is difficult to picture the Planet Earth at this time for volcanic action was occuring continuously and all over the surface of the Planet. Huge amounts of energy exploded from the planet's upper levels of material. In time, the crust began to form on the surface as it began to cool and the plasma-like materials solidified. In the meantime, the large amounts of water which were thrown up by the volcanic action formed heavy clouds which condensed as rain. The heat of the surface was so great, however, that for a long period of time the rain did not reach the surface but vaporized as it fell. This action hastened the development of atmosphere. The action continued and as the cooling occurred pockets of water began to accumulate. During the 600 million years which Gribbin assigned to the formation period of the Planet Earth, the interior as well as the exterior was shaping up to be about what we have now. The heavy iron moved down to become the solid core with a liquid layer of nickel-iron surrounding it. The other layers moved into place, all following the laws of physics. When the surface became solid, the cooling process moved rapidly to achieve what has been calculated to have been the original surface temperature of about 27°C. By this time it could be said that the Planet Earth was ready for development as a place of habitation for plant life.

It is believed that during the turbulent times that followed, rain fell in torrents for several hundred million years! The small pools enlarged to become oceans. The lava flow from the continuous volcanic eruptions spread out to form the ocean floor. In some places the lava output was so great that volcanic peaks rose up above the level of the ocean water and mountains formed. This kind of action is still going on. Another action developed in which the solid crust of the planet surface buckled under the pressure from below and formed high, craggy mountains of rock. This action also brought into being the valleys and plains that we know so well. Rivers began to flow and small lakes appeared.

The entire crust of the planet actually floats on the plasma of the mantle and it is subject to great stress from the actions of the materials seeking to find a way to the surface. Because of this, the surface of the Planet Earth is believed to have changed radically a number of times during the passage of long ages of time, with continents breaking up, moving apart, coming together only to separate again. The movement of such large masses is slow but it can be measured. At present, the North Atlantic is widening at the rate of two centimeters per year. Other areas show the same kind of movement and this has been charted in many places around the globe. A look at any world map or globe will show how South America and the African continent must have at one time been joined.

During this period, the Earth was being prepared for life! An atmosphere was being developed slowly. The volcanos spewed out great volumes of carbon dioxide, nitrogen, other gases and water. The nitrogen was heavy enough so that the gravitational pull of the Earth held it while the lighter gases, such as hydrogen and helium, escaped into the space of the Solar System Sphere. Thus the quantity of nitrogen gradually built up to what it is today—$78+\%$. The water in the warm oceans absorbed carbon dioxide which prevented the build-up of a compound in the newly developing atmosphere that would have had a "Greenhouse" effect in which the heat of the Sun, reflecting from the surface of the planet would be trapped. Had this happened, as it evidently did on the Planet Venus, the temperature would have become so great that the water would have boiled off. Life would then have been impossible on Earth.

A very significant observation should be made at this point. A Planet, to support life as we know it, had to have an abundant supply of water. Thus the temperature would have to be such that it would not boil off the water nor freeze it solid. Planet Earth with an original surface temperature of about 25° to 27°C was about right. Venus, almost identical in size to the Earth, was a little too close to the Sun and Mars was just a little too far from the Sun. Earth was designed for the purpose of supporting life—its distance from the Sun was just right!

The early atmosphere was composed of carbon dioxide, nitrogen, water vapor and other gases in tiny amounts. This atmosphere had built into it a thermostatic control! As the warm waters of

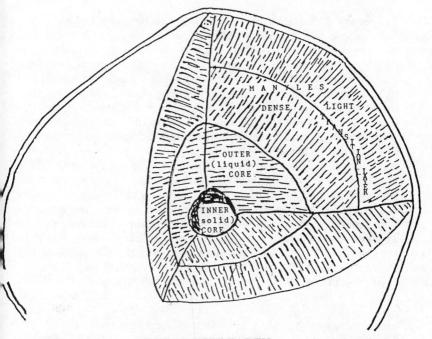

THE PLANET EARTH

Very little is known of the chemical composition of the interior of the Planet Earth. We know about the crust—so thin that it has been compared with the skin of an apple. All across the planet, however, there are active volcanos that work their way up from the plasma below the crust, bringing new material to the surface. As almost everything we have in the crust came into place by way of volcanic action, we can assume that the elements of the crust are the elements of the plasma—molten, we say, to cover our ignorance of the real composition of the mantles that lie between the surface and the outer core of liquid nickel-iron. The contents of the mantles, 1,790 miles from the surface to the outer core—comprise more than 80% of the total volume of the planet. The outer core appears to be a mixture of liquid nickel-iron, 1,380 miles thick. Below this is the solid iron core—780 miles from its edge to the center. The significance of the metalic core is great for it is responsible for the development of the magnetic field that surrounds the planet—forming the magnetosphere which protects the Earth from the deadly Solar Wind. The core is solid because of the great pressure that is exerted on it—the weight of all that lies above.

the newly forming oceans dissolved carbon dioxide out of the atmosphere, the probability of this gas developing a "greenhouse" effect was substantially reduced. The white clouds of water vapor floating above the surface reflected the rays of the Sun—which had a cooling effect. As a result, the average temperature dropped to 15°C where it has remained to this day. Gribbin points out that if the earth should warm up a bit, more water would evaporate from the oceans making for more clouds in the atmosphere which would block out the Sun's rays by reflecting them back into space. If the Sun should cool a bit or if the Earth for some reason should become a bit cooler,there would be less evaporation, few clouds and more heat coming through to the surface of the planet from the rays of the Sun. Either way, the temperature of the Planet Earth would not change radically or disturb the heat range needed to preserve the Earth as a wet Planet, capable of sustaining life. These are not happenstance coincidences. They are in line with all the other items we have considered which indicate that the Creator-God, in His design for life, planned it that way.

A rock planet, with no soil and just the beginning of an atmosphere, was an inhospitable place for life of any kind to develop. But the process of planet-preparation had begun. The lava flow produced substance that in time would become rich aluvial soil. The earthquakes and the buckling of the crust of the Planet produced mountains of granite and other kinds of rock. The clouds of water vapor produced by the constantly erupting volcanos produced not only water but violent rain storms with continuous lightning strikes. As the seasons changed, there were periods of snow and ice. As the atmosphere developed, wind storms were almost constant. All of these forces combined to provide the erosion process that would reduce the rocks and lava to the soil we know and use. A thousand million years passed before the Planet was ready for the first forms of life. Fossil records indicate that primitive plant life forms appeared about 3,000 million years ago. How they came into being, how they developed, is our next point of interest. The scientific evolutionist claims that they just happened to form when the right chemicals, atoms and molecules, happened to meet each other under the right conditions. The laws of probability have been stretched far beyond the limits of credibility.

Paul Davies, on page 65 of *The Runaway Universe*, tells us:

"It is not known precisely how or when life first appeared on Earth. Sometime in the first one-and-a-half billion years after the formation of the planet, one or more of a variety of chemical processes could have triggered the beginning of a long and complicated series of molecular rearrangements ending in the first living, self-replicating organism."

This statement is an expression, in vague and uncertain tones, of the wish of the scientist. There is in it no explanation of "how" the "first living, self-replicating organism" came into being. When we think in terms of the hugeness of the Universe, the space in which the atoms and molecules of matter move in utter isolation—for their movements must be random according to what we have been told—it is a bit difficult to see that it provided a likely setting for the beginning of a chemical process which would produce living organisms. Should such a "miracle" have happened, all on its own, and a "living, self-replicating organism" was formed, we ask the scientist how that organism was powered? As we shall see later on, the scientists hold strongly to the concept that the only power in the Solar System that is usable for living things comes from the Sun by way of the photosynthetic process—which could not have been in existence at the time the first self-replicating molecule is said to have come into being. The photosynthetic process is complex, as we shall see later; it does not happen by chance! The chlorophyl molecule, for instance, is a compound of atoms of carbon, hydrogen, nitrogen and oxygen, surrounding an atom of magnesium and this molecule is only one part of the system that was designed to capture the energy of the Sun for the purpose of providing a usable power source for life—self-replicating molecules included!

Only by the power supplied by the formula E = the Spirit of God + Active in Time could there be the energy available for a self-replicating molecule to function. The sustained expressed will of the Creator-God supplied the energy required when the time came for the first living organism.

"Life began when—somehow, somewhere—a combination of chemical reactions produced a molecule that was capable of making copies of itself by triggering further chemical reactions" (Gribbin, p. 166). From that point, in the same paragraph, he makes a giant leap to state, "So the fundamental molecules of life are today concealed within a protecting wall of materials as individual cells—and many millions of cells function together to make up a human being or a tree."

At this point, we must again enter our objection to the method of the scientific evolutionist—which is certainly not in tune with the "scientific method." We cannot accept the concept that "somehow, somewhere" life began when certain very specific atoms, molecules and compounds in some primordial sea came together totally by happenstance and in exactly the right combinations and molecular forms and brought forth "life". Nor can we accept the facile leap from such "molecules of life" to the protective cell without any explanation as to how this very complex structure came into being. The cell is so complex in its structure that we feel it is necessary to devote a whole chapter to it. The leap from inanimate chemical compounds to living cells is enormous.

The complexity of the chemical compounds, the molecules involved in the "amino-acid" group, is so great that the likelihood or probability of such a happening is so remote that even the billions of years that the scientists have asked for, fades into insignificance. We must recognize that all of the complex molecules and compounds which had to be involved in such an event had also to have been developed by the same random process—atoms colliding by happenstance, molecules forming "willy-nilly", compounds developing by accident. And somehow, somewhere, the right combinations of all the component parts of the life-form had to be in the same microscopic spot under the proper conditions of temperature, pressure and possibly atmosphere and environment. Even though such an assembly of inanimate atoms and molecules should have occurred, the "essence of life" would have been missing. The more closely one examines all the parts of the action that would have had to occur, the more impossible the situation becomes.

The scientist, however, does not appear to be bothered by the problem. Given enough time, he argues, anything can happen! "Quite probably, even amino acids were present in the primeval soup, a revelation which prompted Jim Lovelock to say that 'it seems almost as if our galaxy were a giant warehouse containing the spare parts needed for life' and to draw an intriguing analogy with a planet made up entirely of components of watches. Given a long enough time—perhaps a thousand million years—argues Lovelock, tidal forces and the movement of the wind will assemble at least one working watch. Given a planet rich in the components of life, chemical reactions will produce a replicator molecule within a thousand million years or more." This must be recognized as pure speculation couched in wishful thinking! If the scientist believes that he must rule out the possibility of a guiding hand of a Creator-God, there is no other way for him to go but that of fantasy. Should the fantasy of Lovelock materialize in the form of a complete watch, we suggest that the watch would rust in its case before it measured any time. The missing ingredient would be "The Energy Source" to wind up the spring to apply power to the wheels of the watch! This would also be true for the speculative self-replicating molecule—the assembly of the parts, even in place, is not enough—the spark of life has to be added. Hawking in another setting (*A Brief History of Time*, page 140) makes a comment that touches this point. He says, "With the success of scientific theories in describing events, most people have come to believe that God allows the universe to evolve according to a set of laws and does not intervene in the universe to break these laws. However, the laws do not tell us what the universe should have looked like when it started—*it would still be up to God to wind up the clockwork and choose how to start it off.*"

Having rejected the simplistic solution of the scientific evolutionist with respect to the development of life forms on the Planet Earth, it is our responsibility to present an alternative. To do this, we must go back to the fundamentals. We must review what has already been said about the original energy-particles-waves that composed the Creation Cloud after the Big Bang. We have no problem with the scientist's findings with respect to the composition of the Creation Cloud nor with the time involved in the cooling process which had to pass before the electrons and protons could be joined together in an atom of hydrogen, the first element of matter in the Universe. We have no problem with the continued process of Universe development in which the original cloud broke up into the huge galactic system in which stars are born. We do have some difficulty with the element-formation process which the scientists have described. This does not detract from the main theme of the scientific community other than to raise questions about how all of the natural elements were to be found on the Planet Earth—questions which we hope to resolve before we reach the end of our study.

In the meantime, we have come to the point where the Planet has been formed. The volcanic eruptions, the massive earthquakes, the erosion of the surface rock caused by heat, cold, lightning strikes, wind and water were all at work, preparing soil for the young Planet so that plant life could be developed. Oceans formed, lakes and rivers appeared on the surface of the Earth as its crust moved above the surface of the plasma. Time passed—changes occurred—an atmosphere was developed which was suitable for the first forms of life. Later, as the plants began to cover the Earth, oxygen was put into the atmosphere. It was a by-product of the photosynthetic process used by the plants in the transformation of sunlight (photons) into the energy required for plant growth. This provided an atmosphere that would later allow animal forms of life to develop, using the oxygen discarded by the plants. The animals, in the respiration process, used the glucose of the plants in their life process and discarded carbon dioxide which the plants must have to sustain their life. Thus, each life form supported the other. But, we are getting ahead of our story. We must go back to the fundamentals and work through the process as it was designed and powered by the Creator-God. This we will do in the following chapter.

CHAPTER X
THE BEGINNING OF LIFE

The law of the Conservation of Energy indicates that something cannot come from nothing. The existence of the original energy particles or waves of the Big Bang creation cloud must be explained in terms that do not require a "somehow" in the equation. This can be done when we write Einstein's equation $E = MC^2$ as E = The Spirit of God + Active in Time, wherein E = the Sustained Expressed Will of the Creator-God.

Matter formed in the Universe when electrons orbited or "surrounded" protons to form the hydrogen atoms. The process of development continued when neutrons joined the nucleus of the hydrogen atom to form heavy hydrogen. A big step forward was made when hydrogen nuclei fused to form the helium-4 nucleus. The process moved forward step by step and in accord with the principles laid down in the Laws of Nature until the other elements were created. This was discussed in Chapter IV. As we come now to a discussion of the beginning of life on Planet Earth, we will review what was stated there and remind ourselves that we must look very closely at certain original particles in the Big Bang Creation Cloud: the electrons, protons and neutrons. Without them and their interactions there would be no Universe.

The electron appears to be the fundamental particle—that is, it cannot be broken down to simpler particles. It has a negative electrical charge. It has an energy source that appears to be inexhaustible. It is the active agent in the formation of all the natural elements; it is the active agent, as we shall see later, in the photosynthetic process wherein the chlorophyll in the plant leaf transforms the energy of the Sun into a chemical compound, the ATP molecule which provides a plant with energy for growth and the production of glucose. The electron is the active agent that in the animal cell reverses the photosynthetic process through the respiration process to change the glucose of the plant back into the ATP molecule which then provides the animal cell with energy for growth. A bit later we will describe the ATP molecule.

A stream of electrons passing through a conductor is called "electricity" and is used to provide power to illuminate a dark room with a light bulb and to bring energy to innumerable types of industrial machines, scientific instruments and to Voyager II on its mission in space. The electron is pure energy—it is not a material substance which means that in some sense it is spiritual, like the expressed will of the Creator-God. As its energy is continuous, we say that it is powered by the "sustained expressed will of the Creator-God. This explains its inexhaustible energy; for as long as it continues to be the will of the Creator-God, the electron will continue to function and the Universe will continue to be.

The electron is almost weightless and yet its mass can be calculated. This means that there is an attraction between the electron and the source of its energy—the Creator-God. This attraction we call gravity. It exists universally and operates within a framework of physical law. It ultimately becomes the greatest power in the Universe even though at times it appears to be very weak. This is in keeping with our stated formula: E = the Sustained Expressed Will of the Creator-God. We have used "unit of gravitational force" (ugf) as a term to quantify the value of the gravitational force—in keeping with this concept that gravity represents a measure of the attraction level between an energy form or mass, in whatever combinations, with the Creator-God. As the electron appears to be the fundamental energy form and cannot be divided, we have assigned to it the ugf of 1. This is my terminology (not the scientists') I use it because it helps me understand the meaning of gravity and its significance.

Before going on to discuss the proton it is important that a statement be made on what gravity is. Gravity is another of the mysteries which the scientific community has not been able to explain adequately. The scientists know much about it and can use it in their equations—but to define it, to tell what it is, escapes them. Einstein struggled in vain during the last forty years of his life in an effort to find an answer that would resolve the mystery. John D. Barrows, University Lecturer in Astronomy at the University of Sussex, England and Joseph Silk, Professor of Astronomy at the University of California, Berkley, wrote of this in their book, *The Left Hand of Creation*, page 74, "Einstein devoted the last forty years of his life searching for a description of gravity and electromagnetism that would reveal them to be just different manifestations of a single more basic interaction. His search led nowhere." These men continue, to say, "Gravity is a most unusual force. It stands apart from the other three forces of nature in that no means has yet been found to apply the quantum theory to it nor it to the quantum theory. The strong, electromagnetic and the weak forces are more akin, and are all described by quantum theories." The "strong" force deals with the quarks—bonding them together into hadrons on the way to the development of the proton. The "electromagnetic" force deals with magnetism, electrical charges and the like. The weak" force deals with the decay or breakdown of radioactive matter.

Each of these forces exist to fulfill a specific mission that is in the will of the Creator-God and they have no other function. The force of gravity, on the other hand, acts on all matter; it strips iron atoms of their electrons to begin the fission process in element formation; it is the force that causes stars to implode as they come to the end of their life, reducing them to White Dwarfs and finally Black Dwarfs. Gravity crushes the neutron star and literally forces it out of the Universe. Gravity deals with the whole of the Universe. This kind of power would indicate that it is a force of a different kind as compared with the others—it may embrace them and control them as the superior force operating as the Creator-God's control agent. So we say that it is the direct line of control between the Creator-God and the matter of the Universe. In spite of its enormous power it is also gentle, allowing the leaf to flutter slowly to the ground. The three other forces may proceed from the force of gravity but are not thereby a part of it, as for instance, being ⅓ of it. The force of gravity is in no way diminished by the "working" of any or all of the three inferior forces. It can be said to be the Superior Force and it is responsible directly to the sustained expressed will of the Creator-God. In this sense, if we think in terms of "time", we must remember that gravity as a "force" appeared simultaneously with the out-flow of energy from the Creator-God in the Big Bang explosion. The Laws of Nature were in effect from the beginning wherein the will of the Creator-God became articulate to guide the proper development of matter under the control function of the force of gravity.

Thus, the force of gravity was from the beginning of the Universe the control agent of the Creator-God; and in the split-split second after the Big Bang, it released the other three forces of nature to perform their special functions. It would appear to this observer that the "Unified Force" the scientists look for is to be found in the Force of Gravity, as the Creator-God's Control Agent! This concept appears to be in keeping with the statement of James Trefill, in his essay, "The New Physics and the Universe," as it appears in *The Universe*, edited by Byron Preiss. "Consequently, they tell us that in the moment of birth, the universe was as simple as it could possibly be. There was one sort of particle interacting through a single, fully unified force." (Gravity?)

If you were to put your mind into a state of "free-wheeling" and think about the force of gravity—which operates throughout the entire Universe, acting on mass with a power of attraction, and if you were to eliminate this power—what would happen? You might at first think that the Universe would fly apart for nothing would be held to anything. This, however, would not be the case, for there would be no motion. The apple would not fall to the ground, the planets would not orbit the Sun, the Suns would not orbit their gravitational centers for there would be no gravity, the Universe would be dead. Of course, if there were no gravity—there would not have been a Universe! Thus, the force of gravity is the instrument of the Creator-God in bringing motion into being and thus life to the Universe. The Creator-God used the force of gravity to create the Universe and still uses it in the continued development of it.

Barrow and Silk continue to speak of this "unusual force" and they tell us that the Solar System was developed "under the relentless pull of gravity." They state also that "Gravity is undoubtedly responsible for the formation of galaxies and their unusual distribution in space."

Stephen Hawking in *A Brief History of Time.* states, "Grand unified theories do not include the force of gravity. This does not matter too much, because gravity is such a weak force that its effects can usually be neglected when we are dealing with elementary particles or atoms. However, the fact that it is both long range and always attractive means that its effects all add up. So for a sufficiently large number of matter particles, gravitational forces can dominate over all other forces. *This is why it is gravity that determines the evolution of the Universe.*"

From these statements it would be fair to infer that Barrow and Silk ascribe the development of the Universe—from the moment of the Big Bang—to this unusual force called gravity. Then, almost as though they would defend themselves, they, like other scientists we have quoted before, make plain that the power of the force of gravity must not be thought of as representing anything other than the "natural." No Creator-God is to be implied. They say, "Gravitational theory alone is capable of providing a natural explanation for both the characteristic size and pattern of the galaxy distribution. *No specially initially imprinted scales, patterns, or new forces seem to be required."* (emphasis mine).

We would insert at this point the thinking of Isaac Newton, the genius who first described the Laws of Gravity in the late seventeenth century. We take this information from the book *Coming of Age In The Milky Way,* by Timothy Ferris, author of the best-selling books, *The Red Limit* and *Galaxies.* Professor Ferris teaches science writing and astronomy at the University of California, Berkley.

Ferris, on page 120 of *Coming of Age In the Milky Way,* quotes

Isaac Newton: "The Cause of Gravity is what I do not pretend to know . . . I have not been able to discover the cause of those properties of gravity from phenomena and [so] I frame no hypothesis."

Ferris goes on to state, on page 121, "In any case, the ultimate unsolved questions were for Newton, not scientific but theological. His career had been one long quest for God; his research had spun out of this quest, as if by centrifugal force, but he had no doubt that his science like his theology would redound to the greater glory of the Creator."

"At the conclusion of the *Principia*, Newton asserted that "this most beautiful system of the sun, planets and comets, could only proceed from the counsel and dominion of an intelligent and powerful Being."

On page 122, Ferris again quotes Newton, "The Motions which the Planets now have could not spring from any natural Cause alone, but were impressed by an Intelligent Agent." "In modern scientific terminology," Ferris adds, the question he was addressing is called the problem of initial conditions. We think that the formation of the solar system can be explained in terms of the workings of natural law, but the authorship of the laws remains a mystery. If for every effect there must have been a cause, then what, or who, was responsible for the *first* cause? But to ask such questions is to leave science behind, and to enter the precincts still ruled by Saint Augustine of Hippo and Isaac Newton the theologian."

The failure of the scientific community at this point can be chalked up to its refusal to put the Creator-God into their equations. The Creator-God is responsible—he did not abandon his creation to stumble along its way, as Davies stated. He maintains a constant hold on the Universe. The Sustained Expressed Will of the Creator-God = The Spirit of God + Active In Time—which puts Einstein's equation $E = MC^2$ into a theological setting where it belongs! The mystery of gravity cannot be resolved apart from this relationship to the Creator-God—as Einstein's futile search indicated.

We return to our discussion of the building blocks of matter as they were found in the Big Bang creation cloud. The proton is the nucleus of the atom. It has a positive electrical charge. The negatively charged electron orbits or surrounds the proton to form an atom of hydrogen. The proton determines the nature of the element. It also is an energy form and thus non-material. It has a mass 1836 times that of the electron which means that its attraction to the source of energy is that much greater. We thus give it a ugf of 1836. Only when an electron goes into sustained orbit around a proton does matter come into being. All of the natural elements that make up the Universe are thus basically the same stuff—the number of protons in the nucleus of the atom of each element makes the difference. The configurations of the protons with an equal number of electrons in orbit is not a random matter for as was shown in Chapter IV, the atomic structure for each element is designed very precisely. As the proton is a pure energy form, its existence also is dependent upon the sustained expressed will of the Creator-God.

The proton, however, is not a fundamental particle. It has been broken down into 32 subparticles, each of which plays a vital role in the make-up and function of the proton, as was described in Chapter IV. There is much to be learned about the proton. As this may involve knowledge of pre-Big Bang time, it would be necessary to probe into ETERNAL TIME to get the answers. That very likely will not be possible for anyone living in UNIVERSE TIME. This, however, does not prevent us from using the proton as we know it to understand the Universe and the matter of it.

The fact that the scientists have succeeded in breaking down the proton calls for a careful look at it. As these subparticles are necessary to the formation of a proton and as the scientific evolutionist insists that all happens by the random actions of energy particles, we are again faced with a problem. How were the 32 specific subparticles assembled to form one proton? And as the protons are not selfreplicating and exist in numbers so large that it would be about impossible to number them, how did the 32 particles times the number of protons in the Universe get together in a manner that the formation of protons was the result? How were they brought together in proper sequence to make a proton, all by happenstance? Looked at from this point of view, the position of the scientific evolutionist again becomes untenable. Recognizing, however, that the creation of the protons took place in the Big Bang explosion, we can only accept them as we have learned to know of them—that they are the product of the will of the Creator-God, expressed and sustained. We meet them in the Creation Cloud and know of them as the nuclei of atoms of elements which become real matter only when they enter a sustained relationship with electrons. Their significance is realized when we consider that it is the proton configuration in the atom that determines the nature of an element, whether it is hydrogen, carbon, iron, etc. The fact that both the protons and the electrons are the product of the sustained expressed will of the Creator-God, both being energy forms, and thus of the same substance, it appears that we must recognize that while they may be of the same substance they are programmed differently. They have different missions (functions) as they work together to form matter. We could even say that they represent different phases of the same substance. The neutrons, photons and neutrinos could be put in the same category. Thus, at the very beginning, these energy packets (?) formed. As we go on in our study we may find out more about them.

The neutron is another significant energy particle found in the Creation cloud. It is electrically neutral and it is believed to be an electron and a proton that have been joined. The neutron is unstable and by itself will break down in about fifteen minutes into an electron and a proton. The neutron is neutral electrically—again indicating that the negative electron and the positive proton have cancelled their charges. The weight, (mass), of the neutron is just a bit more than the proton—or, according to our system of counting, it has a 1837 ugf—the weight of a proton and an electron. An important function of the neutron is its capacity to enter the nucleus of an atom of an element, adding weight, or mass, but without changing the nature of the element. As Maffei says, "The number of neutrons in the nucleus of an atom of a given element is not always the same. It is possible, therefore, to have variant atoms of an element with different masses. These variant atoms are called isotopes" (Maffei, p. 25). The neutron's capacity to enter the nucleus of an atom of a given element in different numbers plays an important role in element formation and also in the chemical reactions of polyatomic molecules because of the different masses assigned to the isotopes. While most isotopes are stable, not all are—which means that the unstable isotopes can be radioactive. An unstable isotope of a given element may thus decay or break down and in doing so change the original element into a different element. "But some isotopes of some elements are radioactive, spontaneously emitting particles and being transformed into a different element. Aluminum 26, for example, contains thirteen protons and thirteen neutrons in its nucleus, but is unstable (the stable form, aluminum-27, has one extra neutron)" (Gribbin, p. 78). Gribbin then goes on to describe how aluminum 26 in its decay process releases a positron—a positively charged counterpart to an electron—which converts a proton in the nucleus into a neutron with the result that the element is changed to magnesium-26, with 12 protons and 14 neutrons in its nucleus. Remember, all matter is made of the same stuff! Different configurations of protons in the nucleus makes the difference. Thus the role of the neutron in element formation becomes significant.

We have dealt so extensively with these three energy forms which resulted from the Big Bang because they are the building blocks of the elements of matter. As has been shown in Chapter IV, the building of elements is an orderly process, following a strict code—even though it may be done in violent action. The order and design in the entire procedure is apparent, the purpose is clear, the building of the matter of the Universe leading in time to the Universe as we know it and to life on a Planet Earth.

The assembly of the sub-particles of the proton in the pre-Universe time to form the proton of the Big Bang Cloud was one step in the process. The actual formation of the proton was a second step. The forming of the atom of hydrogen when an electron went into a sustained orbit about the proton marked the beginning of real matter. The building of the second element, helium, by the fusion of the hydrogen nuclei, was a major step. The subsequent building of the 103 elements, however it was done, was all based on the fusion of nuclei up to the element iron and the fission of nuclei from iron on up to uranium. The entire process constitutes an orderly and precise execution of design and purpose, all in accordance with the Laws of Nature as put in place by the Creator-God.

The Universe, the galactic system, the stars and their solar systems, the Planet Earth, the human being gifted with intelligence, the capacity for reflection, for planning, for inquiry, did not come into place without the guidance of the Creator-God. The Sustained Expressed Will of the Creator-God, working through the Laws of Nature as they were set in place, is responsible for it all. Under the concept of "randomness" espoused by the scientific evolutionist, there could have been no design with purpose. There could have been no identifiable "system" or order to the happenstance collisions of energy particles.

In our last chapter we left our discussion of the Planet Earth as it appeared in the early years, ready for the development of a home for life. It was necessary that a suitable atmosphere be developed together with soil and water. The temperature had to be right for both plant and animal life. As the centuries passed while the preparations for life were being completed, the Laws of Nature functioned well. As the first atom of matter, the element hydrogen, is the simplest of all the elements in its composition and as the development of the elements moved from the simple to the more complex, so also when the time came for the first life forms to appear. The Grand Design began with the simplest of structures. But even these, as we shall see later, were not simple. They were profound in that they were made out of very intricate assemblies of organic molecules and set the form, the pattern for the further development of all life forms. The atoms and molecules involved were not self replicating which means that each one had to be made according to the code that had been set.

At the risk of appearing redundant, we must take this procedure step by step from the non-life to the life form. We have learned that the atoms of all the 103 elements are composed of a nucleus of protons surrounded by shells of electrons. The first element, hydrogen, had only one shell with one electron. The shell is capable, however, of having only two electrons. When it acquires a second proton in the nucleus and a second electron on the same shell, the element formed is helium, an inert gas. The next element in line is lithium which has three protons in the nucleus and thus must have three electrons in orbit. As the first shell was "full" with two electrons, the third formed a new shell which surrounded the first. It has a capacity for eight electrons which means that lithium has seven empty electron positions on its outer shell. This also means that lithium will be chemically active with other atoms that do not have a full complement of electrons on their outer shells. This represents a principle that follows through all of the elements. The outer shell can have no more than eight electrons. When it has less, the element will be chemically active, joining with other elements to form new elements or chemical compounds of almost infinite variety. Whenever an element has the outer ring exactly filled with eight electrons, the resulting element will be inert. Thus, the formation of chemical compounds is accomplished in an orderly manner. The science of chemistry is built on this fact.

Now, as we are about to look at the development of the first life form, the molecule, we must realize that the scientific evolutionist claims that the early chemistry happened by sheer happenstance while those who believe that the Creator-God had a part in it are of the opinion that the process was orderly and in accordance with plan.

The chemistry of life is built around just a few of the elements which were among the very first that came into being—all of them being found within the frame of the first two electron shells. The elements are hydrogen, oxygen, carbon and nitrogen. As the Planet Earth was gradually developing the atmosphere, the soil, water and temperature for a life sustaining planet, all of these elements had a role in the process of life building. Nitrogen supplied the bulk of the mass of the atmosphere. It was also involved in the development of the amino acids which were essential to the building of the "life molecule" Carbon and oxygen were combined as a gas, carbon dioxide, an essential ingredient for the first plant life. There was almost no free oxygen in the early atmosphere. Had there been free oxygen in quantity, plant life as we know it could not have developed for the oxygen would have poisoned plant life. Later on, when plant life flourished, it became the source of free oxygen in the atmosphere! Oxygen is a by-product of the photosynthetic process in which plant life captures the energy of the Sun and uses it for its own growth and for the production of glucose. Thus free oxygen was put into the atmosphere by the plants in amounts that were sufficient for animal life to develop and yet not in quantities large enough to be detrimental to plant life. As plants covered the surface of the Planet they released oxygen in quantities that built up the total supply at 20 + % of the volume of the atmosphere.

Interestingly, the build-up of oxygen stopped at that point and has remained constant at that level ever since. As animal life developed later on we find that the by-product of animal life is carbon dioxide, released into the atmosphere through the respiration process, to be used by plants again. A more striking example of design with purpose could hardly be imagined. Plants alone on the Planet would soon have destroyed themselves as the supply of carbon dioxide diminished and as the oxygen in the atmosphere increased. Without free oxygen in the atmosphere there could be no animal life. As it is, each life form supplies the other with the necessary ingredients for life. Hydrogen, among the four primary elements, was essential at all points of the chemical reaction. It combined with oxygen to form water, with carbon to form methane, with carbon and nitrogen to form the amino acids, the precursor molecules of life. The elements of sulphur and magnesium, both found on the third electron shell, were also significant in the chemistry of life.

We have arrived at the point where the next step in the process of life development for the Planet Earth was to take place. "The chemistry of life is essentially the chemistry of carbon, and in particular the chemistry of very long carbon chains with interesting bits and pieces stuck on to their sides. But—the big question (is) what exactly do we mean by 'life'? The simplest and best way to distinguish between the living and the nonliving is that the living things are able to reproduce—they can make copies of themselves. . . . *Life begins when—somehow, somewhere—a combination of chemical reactions produced a molecule that was capable of making copies of itself by triggering further chemical reactions" (Gribbin, p. 166)*.

The scientist, who takes pride in the "scientific method" cannot be allowed to get by with such an unscientific statement! It takes some doing to develop the "very long carbon chains" that are a necessity in the development of the most simple life forms. Gribbin has supplied a number of diagrams of carbon molecules indicating the importance of carbon in the chemistry of life. We will use one of these to help make our next point. A molecule of methane gas, one of the simple carbon molecules, shows a carbon atom bonded to four hydrogen atoms. Bonded, because the four empty electron spaces on the outer shell of the carbon atom need to be filled to make a stable compound. It could also be said that the four electrons on the outer shell of the carbon atom need to take on four electrons to make the carbon outer shell complete with eight electrons. Methane is just a simple carbon compound, a long way from the very complex amino acid molecules that had to be developed before a "life molecule" could be assembled. Consider now the problem the scientist faced when he had to insist that not only the methane molecule but all the long series of ever more complex carbon compounds leading up to the "one" were formed by happenstance collisions of atoms of carbon and hydrogen, complicated in the case of amino acids by the introduction of nitrogen atoms! We can understand that it was easier to say simply that "somehow, somewhere" it happened! The width of one atom of hydrogen is only about one 250,000,000§/ of an inch, as we learned in Chapter IV. Nevertheless, these specific hydrogen atoms that we are dealing with in this illustration, tiny as they are, had all four to collide with this specific carbon atom in such a manner as would allow them to bond together to form a molecule of methane gas. To have this happen once would appear to be a very unlikely random act but as methane molecules are not self-replicating, it would have to happen innumerable times to form the abundant quantities of methane gas on Planet Earth. Methane gas is what we commonly call "swamp gas" and there is a lot of it! Thus, to move from this simple carbon molecule to the highly complex carbon chain molecule with "interesting bits and pieces" attached, and to do this all by happenstance collisions of atoms, reaches beyond the limits of possibility.

In the carbon chemistry leading to the formation of the molecules which would "come alive" there are increasingly complex compounds. The more complex they are, the less likely would be their chance of being formed by random movements of atoms and molecules. The amino acids, which are extremely complex, represent about the last step before the appearance of the "life" molecule. One gets the impression on reading Gribbin (*Genesis*, p. 165) that he is somewhat desperate in his defensiveness when he states: "This doesn't mean that some guiding hand stuck the chemical bits together in the primeval soup. As soon as molecules that could reproduce themselves—living molecules—appeared, (!) *the reactions they are involved in were no longer random.*" The emphasis is mine. He only emphasizes the weakness of his position by insisting that these "living" molecules simply "appeared" without any help.Then, "out of the blue," he abandons the concept of random development.

We are not saying that the hand of the Creator-God actually put the four hydrogen atoms in place with the carbon atom to form the methane molecule. Rather, we are saying that the Creator-God set in place the design for all of the Universe, from the development of the elements to the development of the life forms. It was not a "somehow" operation but one of Law. "The belief that (such) laws of physics exist and that they are capable of being discovered by rational inquiry is basic to all science." So says Thomas L. Swihart in *Journey Through the Universe*, (p. 210). The scientists can perceive and make use of the Laws. They do not seem to be willing to accept as fact that the Law did not come into being by happenstance. Were it all by randomness, there would be no order in the Universe and it is to be doubted that any Universe would exist. Should one exist under such circumstances, scientists, if there were any, would be unable to function for there would be no laws, no equations to go by!

Thus, the elements developed as they did because it was the will of the Creator-God. The polyatomic structures, molecules and compounds, also developed as they did because it was the will of the Creator-God. Once the original life form had been created, under the will of the Creator-God, established within the environment which had been prepared in advance, the continued development of life was assured, for it was the will of the Creator-God that it should be so. The inanimate atoms and molecules did not have "life" in and of themselves. The so-called "life molecule" that was formed under the design of the Creator-God was to be the receptacle of the "life" that would be given to it when all was ready. It could be said that the chemical molecule provided the environment necessary to protect the life, the spirit, the soul—whatever it may be called. No less a scientists than Charles Darwin stated that it was the Creator who "breathed life into the first form."

We must not get hung up on semantics. Life is a "given". The structure of the life molecule provided the body for the "living spirit" which controls all the parts of the living thing—plant or animal! There is a quality of "aliveness" that is not inherent in the molecules of matter that go into the make-up of a body. When that quality leaves the body, there is no life in what is left. In time, the body reverts to the elements from which it was made. This concept the scientists have simply glossed over as they write of the amino acids, the necessary building blocks of life. Building blocks we can agree to, but that the "aliveness" is derived from the particular arrangements of the atoms of the elements in the building blocks does not follow. The "aliveness" comes from the specific sustained expression of the will of the Creator-God. The Bible tells us that the Creator-God "breathed the spirit of life" into the body that had been formed (Gėn. 2:7). This would be true for all life forms, plant and animal. The life-spirit is a given, coming from the "outside." It has an existence apart from the limits of the body, the form. The so-called self-replicating molecule of the scientific evolutionist implies the presence or existence of an Intelligence to direct the replication process—the assembly of the required atoms and chemical compounds. This "presence" is not inherent in the lifeless atoms of the molecule; it is the "given" which has been and is supplied by the will of the Creator-God. The energy source that is necessary for life is not to be found within the molecule. There is only one source of life-sustaining energy in the Solar System—the Sun. The heat of the Sun, however, would kill all forms of life if it were to be applied directly. The design of the Creator-God provided for a process whereby the Sun's energy would be captured and translated into chemical energy in a molecule called ATP. It is a product of the photosynthetic process which takes place in the green leaves of the plant.

In the meantime, we must realize that the self-replicating molecule of the scientific evolutionist could have no possibility of living or reproducing until plant life and the photosynthetic process was in operation. Thus, the first "self-replicating molecule" must have been the simplest of all plant forms. In the following chapter we discuss the photosynthetic process in some detail and and we ask the reader to think of how such a complex system could be put together by happenstance, and of how the body of the first life form, with all of its parts having special functions, could have been formed by chance. It would be far more reasonable to say that when the Creator-God had supplied the life form, with the equipment necessary to sustain life he also breathed the essence of life into the form.

The continuous movement toward a living molecule—a plant form—kept on until the time when the Creator-God said to his workmen, "Let there be!" While mystery surrounds them because we know so little about them, we must recognize that the hosts of angels of which the Bible speaks are always about, carrying out the will of the Creator-God. How they may have functioned in this early Solar System time, we cannot say. They were, however, pre-Big Bang—meaning that they were with the Creator-

God in eternal time. Theologians would do well to concentrate efforts on learning more about these "intelligent energy forms" for which we have no other name than "angel."

We do not know who wrote Proverbs 8:22–31, nor can we say who first spoke the words that are there recorded by King David about a thousand years before the time of Christ. They cover a great expanse of time and events of great significance, from before the Big Bang to some time after mankind inhabited the Planet Earth. Scientists as well as theologians should study these words:

> The Lord God formed me in the beginning of his way, before his works of old. *I was set up from everlasting, from the beginning, before the earth was.* When there were no depths, I was brought forth, when there were no fountains abounding with water. Before the mountains were settled, before the hills, was I brought forth; while as yet he had not made the earth, nor the fields, nor the beginning of the dust of the world. *When he established the heavens, I was there*; when he set a circle upon the face of the deep, when he made firm the skies above, when the foundations of the deep became strong, when he gave to the sea its bound, that the waters should not transgress his commandment. *When he marked out the foundation of the earth, then I was by him as a master workman*; and I was daily his delight, rejoicing always before him, rejoicing in his habitable earth; and my delight was with the sons of men.

The writers of the New Testament identify "the master workman" of this statement with Jesus Christ, as we see in these quotations from the Gospel of St. John and Paul's Letter to the Colossians:

> "In the beginning was the Word, and the Word was with God, and the Word was God. The same was in the beginning with God. All things were made by him and without him was not anything made that was made. (*John* 1:1–3)
> "For by him were all things created, that are in the heaven, and that are in the earth, visible and invisible, whether they be thrones, or dominions, or principalities,

or powers: all things were created by him, and for him:
And he is before all things, and by him all things consist."
(*Col.* 1:16, 17)

The "master workman", Jesus Christ, who appears to have been in charge of bringing-into-being all of our Universe, was joined by other workmen of the Creator-God, to whom we give the name angels. The fuzziness of the religious concept of angels as portrayed by the writings of the medieval churchmen and illustrated by the colorful and imaginative portraits of chubby-cherubs-with-wings should be swept aside. Serious study should be made of these "spirits" who filled the Old Testament with their comings and goings and who were constantly at the command of Jesus in New Testament times.

Thus, when the time came for the first life forms—whatever they may have been—the Creator-God had a part in providing to the inanimate molecules the living "spirit". Life, as we have said, is a given, for there is nothing in the molecules of carbon, nitrogen, oxygen and hydrogen that can be "alive" with what we call spirit or soul or self. When the spirit leaves the body, life is gone, the body is dead; the cells have no life and the substance of the cells decays, returning to the elements of carbon, nitrogen, oxygen and hydrogen. The bestowal of "aliveness" upon the "life molecule" was another step in the gradual fulfillment of the Design for Life which had been set up in eternal time. The principle, mentioned earlier in this chapter, that life forms began with the simplest and developed gradually into the complex, took over as life on Earth began.

At this point, we should look at the Bible record of the beginning as told in Genesis 1. We stated early on that we do not think of the Bible as a scientific textbook. We do believe, however, that the message it holds was in some manner "given" to the writers. Certainly it would be very difficult to explain on any other basis. Whoever put into writing the first chapter of Genesis must have had an unusual source of information! Consider the fact that it was written thousands of years ago when by the reckonings of modern scientists, mankind was still in its infancy. Yet the words speak with clarity and tell of events happening in a specific order that is in line with our modern thinking and the findings of the scientists. Consider also that the message had no doubt been passed down from generation to generation for a long time before a written language form had been developed. The level of civilization at that time was primitive; nomadic tribes of unlettered people constituted the bulk of the population.

The specifics of the first chapter of Genesis are astounding. Both the scientists and the theologians would do well to study very carefully the data provided. The information given relative to the beginnings of life on the Planet Earth is exactly consistant with that laid out by the scientists. The language is that of the man who wrote, telling the story of the beginning of this planet as it had been told to him by his father or priest, who had heard it by his father or priest. It should not be a point of conflict that some items may not fit precisely. Rather, it is source of wonder and amazement that the basic message is so clearly in line with what the scientific evolutionist tells us was the order of events in the beginning time.

Consider the following items:

1. *In the beginning God created the heavens and the earth.* It could well be argued that the first verse of Genesis included the beginning of the Universe by means of the Big Bang—the creation of the super cloud of energy, the protons, electrons, neutrons, photons and neutrinos. We choose not to push that point even though we recognize it as a valid possibility. The writer was telescoping vast amounts of time in his very brief account of the beginning of the world as we know it. No date is indicated for that event. Certainly, the important word is that the Creator-God brought into being not only the Planet Earth but also the "heavens," whatever may be the true meaning of that word. The writer of Psalm 102 says: "Of old thou didst lay the foundation of the earth; and the heavens are the work of thy hands. They shall perish but thou shalt endure; thy years shall have no end." The eternal nature of the Creator-God was a clearly established concept in the minds of the Old Testament writers. In our day, we must face the fact of Universe time as being our time, while eternal time is the Creator-God's. The Universe will have an end, just as our scientists say, but the Creator-God will continue to be.

2. *And the earth was waste and void; and darkness was upon the face of the deep, and the Spirit of God moved upon the face of the waters.* An excellent picture of the newly formed planet—bathed

in the primeval darkness, empty of life, a desolate sphere, except that the Spirit of God was present. Isaiah wrote: "For thus saith Jehovah that created the heavens, the God that formed the earth and made it, that established it and created it not in vain, *that formed it to be inhabited*: I am Jehovah; there is none else" (45:18). The prophet, and those in the prophetic line long before him, knew that the earth was empty and void in the beginning. There was purpose in the creation of the planet: it was designed to be inhabited, life was to be developed. Stephen Hawking in *A Brief History of Time* (p. 127) states: "It would be very difficult to explain why the universe should have begun in just this way, except as the act of a God who intended to create beings like us."

The darkness in the beginning was the primeval darkness for the light of the newly forming Sun had not yet become strong enough to break through the thick cloud of dust and gases surrounding the young Planet Earth. We must remember that the Sun and the Planets were all forming at the same time, as the scientists tell us. All were coming into being out of the Solar System Cloud which had separated out from the arm of the Milky Way Galaxy. The influence each body had on the others was minimal during this formation period, to which Gribbin has assigned the time of about 600 million years. Gradually, as time passed, the influence of the Sun became dominant. As it became larger and hotter, it took control over the the other bodies in the Solar System. This leads us to the next statement in the Genesis account.

3. *And God said, Let there be light: and there was light.* This was not the light of the Sun as we know it, for the Sun had not yet developed such power. It was the light of the pale embryo Sun as it filtered through the clouds of dust and gas which must have filled the space around the forming planets—not only the Earth, but also Mercury, Venus and Mars. The outer planets were likely too far distant for this early light to reach them. The Planet Earth must have looked grotesque in this pale light as winds began to blow across craggy mountains and the forming bodies of water for with the formation of atmosphere came also the wind which likely blew constantly.

4. Verses four through ten deal with the events in the period of

time during which the sterile Planet was being prepared for life. In this period of hundreds of millions of years, the atmosphere was developed to about what we have today. Erosion produced soil, waters were gathered in oceans and lakes, rivers began to flow across the land areas draining water into basins. The continents in their original shapes were defined as areas of dry land. During this period, as the light of the Sun was still diffused and pale, the darkness of evening and the light of dawn marked the passing of each day. No Sun was etched in the skies but like on a dark cloudy day now, the light was there. Thus, the period of a day was defined as one turning of the new Planet on its axis. We must not be caught up in a fruitless discussion of the meaning of the Hebrew word *yom*. It means what it says: a normal day! But it must also be said that the writer of the first chapter of Genesis was compressing the events of hundreds of millions of years into a framework of words that readers could grasp. The days were days, but how many there were between verses five and eight, no one knows but the Creator-God.

5. *And God said, Let the earth put forth grass, herbs yielding seed; and it was so*. The verses 11–13 mark the beginning of the first life form, the beginning of life. Here the Creator-God breathed life into the first simple plant forms, providing them with the essence of life. We should note here that the Hebrew verb *hiyah* does not indicate a "creative act," as is the case when in the first verse we are told that "God created the heavens and the earth." There, the verb *bara* is used, which has the meaning "to create." Verb forms like *hiyah*, used to tell of the coming into being of the life forms, are of a different kind, indicating that what was developing came out of what had been originally created—"Let there be," "Let the earth put forth," "God made (out of existing substance)"—all of which supports the concept that what we have in this world developed out of what the Creator-God put into the Universe in the beginning. This can be called evolution—a Creator-God guided process, not a random, willy-nilly procedure wherein human beings are finally developed out of a series of "bizarre mistakes," in Gribbin's phrase.

6. *And God said, Let there be lights in the firmament of heaven to divide the day from the night—and there was evening and there*

was morning, a fourth day. The Sun by this time had become hot. The solar wind was blowing out from its surface, sweeping the space of the Solar System clear of the dust and debris of the formation period. The time had arrived for the Sun to be the power that it now is. As the dust clouds were swept away, the full light of the Sun was available, with all of its warmth and energy for the new life forms that were about to appear. The Moon also appeared—reflecting the light of the Sun. The stars in all their sparkling glory were shining in the heavens (Genesis 1:14–19).

7. *And God said, Let the waters swarm with swarms of living creatures, and let birds fly above the earth in the open firmaments of heaven.* The use of the word "firmament" is an example of poor translation. The Hebrew word actually translates as "expanse" or "space." Again, the scientific records indicate that the first forms of animal life were developed in the waters—the primordial swamps, the oceans, shallow and warm. The plants soon covered the earth; oxygen was being fed into the atmosphere and soon was absorbed in the waters. Animal life, even though very primitive, could survive. By animal life, we mean the non-plant life, the small one-cell type that first formed in the shallow, warm waters. Following this development came the more complex forms, the worms, the reptiles, the fish and the birds—all in the same order as the scientists have stated (Genesis 1:20–23).

8. *And God said, Let the earth bring forth living creatures after their kind, cattle, and creeping things, and beasts of the earth after their kind: and it was so.* This passage indicates the continuation of the orderly process of bringing life as we know it into being. The Creator God's evolutionary process was proceeding according to design and purpose (Genesis 1:24–25).

9. *And God said, Let us make man in our image, after our likeness and let them have dominion over the fish of the sea and over the birds of the heavens and over the cattle and over all the earth, and every creeping thing that creepeth upon the earth. And God created man in his own image, in the image of God created he him: male and female created he them.* In this statement there is a significant change. It indicates a counseling between the Creator-God and at least one other Person—who must have been of equal status,

for it is stated, "Let us make man in our image" (Genesis 1: 26–27). We are reminded of the passage from Proverbs quoted earlier: "The Lord God formed me in the beginning of his way, before his works of old. *I was set up from everlasting, from the beginning before the earth was.*" It also reminds us of the Gospel of St. John: "In the beginning was the Word, and the Word was with God, and the Word was God. The same was in the beginning with God. All things were made through him, and without him was not anything made that hath been made."

Further, there is a difference between the development of the other life forms, such as plants, sea life, birds, animals—where the words are: "Let the earth bring forth," "Let the waters swarm"—and this instance where a person is to be formed. Here the record has God saying, "Let us make man,"indicating that this was to be a special act on the part of the Creator-God. Man—male and female—was to be the crown of the creation, being made by the Hand of God, in the image and likeness of the Triune God. To this new creation was given the responsibility to have dominion over all the rest of creation.

The final verses of this strange chapter tell of the commissioning of man, charging him with the great responsibilities that came with the gift of being created in the image, the likeness of the Creator-God. God's blessing on all of the creation is given and the next few verses—which likely should have been included in the first chapter—close out the creative period with the words: *And the heavens and the earth were finished and all the host of them. And on the seventh day God finished his work which he hath made. And God blessed the seventh day and hallowed it: because that in it he rested from all his work which God had created and made* (Genesis 2:1,2).

It is a remarkable chapter, written so long before the scientists of the world had come to study the Universe and the making of it. The words of this chapter cannot simply be brushed aside; they must be accounted for. In all the basic points, the events are in accord with the findings of the scientists. It is a pity that the scientific community feels compelled to insist that no guiding hand, no intelligence, had a part in the bringing of the Universe into being. Jastrow puts it well: "Perhaps the appearance of life on the earth is a miracle. Scientists are reluctant to accept that view, but their choices are limited; either life was created by the will of a being outside the grasp of scientific understanding, or it evolved on our planet spontaneously, through chemical reaction occuring in nonliving matter lying on the surface of the planet" (*Until the Sun Dies*, p. 62).

The agnostic position, that the mind of man must be able to understand and explain the being of the Creator-God before it can be accepted, is the villain in the matter. As long as the concept of a supreme being is made subject to the scrutiny of man, being reduced thereby to something less than man, there is no way out for the scientist. He is forced to grasp "in faith" a concept of the origin of all things that lies completely out of the range of the scientific method. To say that life evolved on our planet by spontaneous generation, totally by happenstance, to say that our planet also is the result of a series of random collisions of atoms which themselves came into being "willy-nilly, to say that the Universe with all of its Galaxies, simply developed out of nothing, is to say that nothing exists, for nothing can come out of nothing! To say such things is absurd. I am as I sit here, coffee on the table, writing these words, thinking these thoughts. I exist—and about me are beautiful people, flowers, birds flying over me, the Sun shining by day and the Moon by night. It is all real—and the Creator-God, together with all his workmen, is responsible for it all. For me or for the scientist to deny the reality because our small minds cannot understand or comprehend the meaning of eternal time and the One who holds it in his hand, is the ultimate defiance.

CHAPTER XI
LIFE AS WE KNOW IT

We have followed the development of the Universe from the time of the Big Bang of creation until the beginning of life on the Planet Earth. We have taken the stand that it all happened in accord with the Grand Design of the Creator-God. We have sought to show why the position of the scientific evolutionist, that it all just happened to happen, is untenable. We come now to a discussion of the development of life from the first simple form to the present—with plant and animal life in highly developed forms flourishing on the earth.

Unfortunately, we find that we must continue to deal with the same problem. The scientist claims that life evolved from the first simple form to the present complex forms by accident, by errors in the transmission of genetic information through the genes. "Changes only happen by mistake, and very few of the copying mistakes are beneficial. . . . Very occasionally, however, an imperfect copy turns out to be better at the job of converting chemical food into replicas of itself than the original—and such rare mutations not only survive but spread through the environment. Over many millions of years, the accumulation of such rare beneficial copying errors gives rise to species as diverse as a mouse and a mushroom. *But the process happens willy-nilly* . . . So the fundamental molecules of life are today concealed within a protecting wall of materials as individual cells—and many millions of cells function together to make up a human being or a tree" (Gribbin, p. 166.)

At this point we must again enter our objection to the method of the scientist, which certainly is not in tune with the scientific method. We cannot accept the "somehow, somewhere" concept that provides for the beginning and development of life by the accidental, happenstance, random chemical reactions resulting from atoms and molecules coming together in just the right time with just the right forms and combinations. These combinations are supposed to have developed not only new and amazingly complex chemical structures but some with the ability to direct the reproduction of themselve! In addition, we cannot accept the facile leap from such "molecules of life" to "cells." The leap is too great and the concept is too simplistic to be credible. Thus, the subject of this chapter will be "The Cell," its make-up and function.

The "life molecule" of the scientist had to have a "home" or a "body" or it would not have functioned as a self-replicating molecule. The cell is a very complex body and it could not have "just appeared," to be available to provide the life molecule with a "living environment." As we go on to discuss the cell, we leave it to the reader to judge whether or not the scientific position is justified in having the cell to simply appear, with no explanation as to how it happened.

John Pfeiffer, author of the Life Science Library volume, *The*

Cell, states on page 9: "There is no life without cells. And just as life itself is diverse, so are the forms and functions of the cells that constitute life. Some cells live alone, as free-moving, independent creatures, some belong to loosely organized communities which move from place to place and some spend their life-time in fixed immobility as part of the tissue of a larger organism. Whatever its form, however it behaves, the cell is the basic unit of all living matter. In the cell, nature had enclosed in a microscopic package all the parts and processes necessary to the survival of life in an ever-changing world."

Before we go into the detailed study of the cell, it would be well to state briefly that the cell—the actual body in which the life molecule was to live—must be accounted for. A cell has a skin, a membrane which incloses the contents, the working parts of the cell. This membrane had to be developed *before* the self replicating molecule of the scientist, for it was inside this enclosure, this space, that the molecule was to function. How did the membrane come into being? According to the scientist, if we hold him to his chosen "turf," the random development of an organic molecule had to be the first state. But one molecule does not a membrane make! Great numbers of these molecules had to form by happenstance. And these had to be assembled at a specific spot where the cell was to be formed. There had to be an agreement (?) to combine to form the membrane; a shape had to be decided upon. If the cell were to function, pores had to be formed in the outer membrane to allow not only for the entrance of the "life molecule" but also other substances to provide food for the molecule and the other organelles within the cell. (Where did they come from? And how?)

"The chemical substances which direct the development of a cell into a specialized unit would be of little use if cells of like character did not organize themselves into tissue. One of the most amazing features of embryonic development is that the right cells almost invariably manage to find their way to the right places in the embryo at the right time. The appropriate cells always seem to be available when interactions involving their development are scheduled to occur. *To date, no one knows how this is accomplished*" (Pfeiffer, p. 104). The same could be said of the entire process of matter development as it moves through complex inorganic-organic molecular development on to the production of the so-called self-replicating life molecule. One thing appears to this writer to be certain: it does not occur by a "willy-nilly" process!

This introduces us to the mysteries of the cell—which we will now begin to study. In doing this, we will describe briefly the composition of the cell, its parts and, in a sense, its organization. We will find it difficult to believe that a cell, so small that it cannot be seen except with a powerful microscope, has many individual working structures with thousands of moving parts! It would be even more difficult to believe that such a complex body could have come into being without the help of an outside Intelligence, the guiding hand of the Creator-God. After this presentation, we will describe the functions of the most significant parts of the cell, all of which had to be put in place before an operational cell could accept a "life molecule."

The most efficient manufacturing plant known to man is the living cell, plant or animal. The typical cell in the human body—and this is quite generally true for all life forms—is very small. If you were to take a lead pencil and make a line on a sheet of paper, the width of the line would be about 100 times the width of the typical cell. This means that without great magnification, you could not see a cell. There are cells so small that their width measures only one-two hundred and fifty thousandth of an inch. On the other hand, the ostrich egg, which is a single cell, is as large as an orange.

There are about sixty trillion (60,000,000,000,000) cells in the body of the average adult human. 50,000,000 cells in the human body die every second, but strangely, 50,000,000 new cells are born each second! How this kind of balance is maintained, how the reproduction system is managed, is not adequately understood. This subject deserves a great deal of study for the dying cells are located throughout the body and are carried out as waste while the replacement materials appear in the proper places at the right time. If biologists could determine how this happens, possibly the cause of death by cancer and other cell-related diseases could be determined and preventive measures developed.

In spite of its microscopic size, the cell is remarkably well organized. It is of great interest that the cell composition of both plant and animal life is almost identical. In this, we see again the efficiency of the Creator-God. All elements are made of the same stuff, electrons and protons, only arranged differently. All cells, plant and animal, have almost identical structures, the difference being in those "devices" which make a plant a plant and an animal an animal. "All cells are built according to a fundamental design which provides them with certain common features apparently necessary for life. Every cell has an outer wall which makes it a room (the Latin meaning of 'cell'). Within the surrounding membrane is a semifluid material called cytoplasm in which the life activities of the cell are carried on. At the heart of the cell is the nucleus, a control center that bears within it the cell's hereditary material, ensuring the survival of the line" (Pfeiffer, p. 10).

The plant cell, as compared with the animal cell, has one very significant part, without which there could be no life on Earth. It has, floating in the cytoplasm (the fluid of the cell) bodies which are called *chloroplasts*, which contain a complex chemical called *chlorophyll*. This substance has the ability to absorb energy directly from the rays of the Sun and to tranform it into sugar, as glucose. In this *photosynthetic process*, the energy necessary for the manufacturing units of the plant cell is provided. The energy is held in a molecule that is developed in the process, called *adenosine triphosphate*, or ATP.

Animals do not have the chloroplasts in their cells and thus cannot absorb energy directly from the Sun for use in the life process. This means that animal life could not exist without plant life, for there would be no energy source available. Animals must consume either plants or other animals which have consumed plants in order to obtain the energy necessary for life. This means that plant life must have been developed prior to the appearance of animal life—exactly what we have observed in the last chapter. It also means that the animal cell must have an almost equally ingenious system to enable it to translate the glucose of the plant to the energy form required for its life. The *respiration process* in the animal cell converts the glucose to the same energy molecule that powers the manufacturing process in plant life: the ATP molecule. Both of these systems will be discussed later in this chapter.

As we continue the discussion of the cell, we must keep in mind that the self-replicating life molecule of the scientist, at the point of its beginning, did not account for more than itself. The molecules of matter that made up the cell were not self-replicating and thus must have been developed by happenstance, if the view of the scientific evolutionist is to be held.

The membrane surrounding the contents of the cell has a thickness of only three millionths of an inch. It has a number of pores, openings which allow atoms and molecules to enter and waste products to exit. There are a number of bodies or structures called organelles within the cell, each having a specific function. We will now briefly discuss each of the cell's components.

1. The *membrane*, or skin of the cell, consists of a thin layer of protein, a layer of fat and another layer of protein. The fat and protein molecules are very complex compounds developed through the carbon chemistry process mentioned in the last chapter. Each of these layers is only one millionth of an inch thick. The probability of such a membrane with its very specific structure developing by chance goes beyond the limit of possibility. Consider that the organic molecules of protein and fat would have to form first—by happenstance—and then they would have to be organized and fashioned into separate layers and bonded together to form the membrane, and this is not all, for the membrane has:

2. *Pores*, openings which allow atoms and molecules to enter the cell. Further, we are told by the scientists, these pores have the ability to select or reject substances, choosing only those that will be productive in the manufacturing processes which take place within the cell. How the pores come to have such powers is not explained. The pores are not alone in this function, for on the surface of the membrane around each pore opening are great numbers of:

3. *Microvilli*, finger-like projections which assist in the process of selecting just the right substances for entrance into the cell through the pores. In their constant waving action, they help some desired "cell foods" to enter and they deter the undesirable. How this is done by inanimate hair-like structures is not known. We say that the design and function of the membrane, the pores and the microvilli are evidence of design and purpose, not happenstance events.

4. The cell is filled with a viscous fluid called *cytoplasm* in which

all the organelles float or move as they carry out their specific functions. This fluid, very complex chemically, must also have developed spontaneously out of carbon compounds—unless the "workmen" of the Creator-God had a hand in the matter! The same must be said for the creation of the organelles—the working parts of the cell—for they too are composed of complex organic molecules. The cytoplasm not only provides an "atmosphere" for the structures within the cell but it performs a very specific function in breaking down the complex glucose molecules and other substances so that they may be used in the various processes taking place within the cell.

5. At about the center of the cell is the *nucleus*, which could be called the "control center," for it supervises almost all that is done within the cell. It is a round body with a porepocked outer covering through which it sends out its messages to the working units of the cell. The mucous-like substance of the nucleus is called *chromatin* in which are the *chromosomes* which contain the cell's hereditary material in long threads that form a twisted ladder-like structure.

6. This structure consists of *dioxyribonucleic acid*, or DNA. It is this molecule that actually carries the genetic information necessary for the reproduction of cells, the replicating process. A diagram of the DNA molecule should rule out the possibility that such a complex polyatomic structure could develop by happenstance. How it could also have within itself the genetic information necessary to direct the reproduction of itself together with the capacity to impart this information to new molecules in new cells is a subject the scientific evolutionist appears to avoid in the readings that I have done. It seems to be just another of the things that just happened to develop over an immense period of time according to the theory of the scientists. More will be said about the DNA molecule later in this chapter.

7. Within the nucleus are some tiny balls, similar in shape to the nucleus, which are called *nucleoli*. These structures appear to function primarily in directing the protein-making process. The nucleoli contains ribonucleic acid, or *RNA*, a substance that appears to function as the message-carrying body for the nucleus.

How the messages are formulated and how they are dispatched is not explained. Again, more on the RNA a bit later.

8. The protein-making is done by *ribososmes*, tiny bodies less than one millionth of an inch thick, believed to be made in or on the surface of the nucleus. The various kinds of proteins—and there are a great number of them—are made up of combinations of atoms of carbon, hydrogen, nitrogen, oxygen and sulphur. There are thousands of ribosomes in each cell. How the proteins are made is not understood. Neither is it known how the ribosomes are made!

9. Within the cytoplasm there is also a "maze of segmented microtunnels" which are called *endoplasmic reticulum*, or *ER*. It is believed that the ER constitutes a "transport system designed to carry materials from one part of the cell to another". It is connected to the nucleus. The ribosomes work closely with the ER, often being attached to it.

10. The ER also brings protein molecules to another structure found within the cell called the *Golgi Bodies*. They appear to be a "packaging agent" for they are sack-like structures which store protein molecules. When there is a need at some point for protein, a portion of a Golgi Body breaks off, to be carried by the ER to the point of need. How these organelles, the working parts of the cell were made is not known. I suspect they are the work of the Creator-God and his workmen!

What we have seen so far of the cell and its functioning parts, indicates a remarkably well organized body—but much more remains to be looked at. We have come to the point where we will look at the bodies of the cell that make a plant a plant, that absorb the energy of the Sun and transform it into a chemical compound which is the power source which enables the cell to function. We are again faced with mystery, for if this function were not in operation, none of the parts of the cell we have just discussed could be operative. A ribosome, for instance, could not be engaged in making protein molecules without the power that comes from the Sun. The nucleus, housing the DNA molecule, could not be ordering the functions of the parts of the cell unless it had power.

It is difficult to believe that all these parts of a cell were somehow set up, formed and ready for action, just waiting for the power to be turned on! It is equally difficult to believe that the power making process developed by itself and was in a sense "standing by" to provide energy to the cell and its parts when needed. It would be far more reasonable to say that an outside Intelligence was at work, putting in place the cell which would have within it the structure that would provide the power when all had been made ready. This, of course, would mean that the Creator-God was responsible for making the cell. As we look at the power-making structure and the processes involved this solution to the mystery will, I am sure, appear to be correct.

Early on in this chapter we mentioned that there was a significant difference between the plant and animal cells at one point. They have different "powerhouses". As we have stated, the plant cell alone has the ability to absorb the energy that flows from the Sun and to transform it into usable energy for life. The power house of the plant is called the *chloroplast*, a structure floating in the cytoplasm of the cell. It is about five thousandth of an inch long. The entire existence of all life forms on earth is dependent upon the functions of the chloroplasts. The process involved is called *photosynthesis* in which the green plant captures the energy of the Sun directly as the rays strike the *chlorophyll* molecules within the chloroplasts. The process involved requires a supply of carbon dioxide and water which the plant draws out of its environment.

The chloroplast is almost an architectural structure consisting of rectangular bodies arranged in a precise order. The entire process of translating the energy of the Sun into a chemical-energy compound calls for a very complex series of events—all of which must be timed exactly. The structures involved and a number of the steps within the process are not yet understood by the scientific community. It is known, however, that out of the process comes glucose and free oxygen, both of which are necessary for animal life. In addition, of course, the plant is supplied with the energy required for its life and growth.

For now, an outline can be supplied. Within the chloroplasts are great numbers of columns called *grana*, which consist of round, flat bodies lying on top of each other to form a column—about a dozen to each grana. The flat disk-shaped bodies are called *quantasomes*. They can be seen only when magnified 90,000 times by an electron microscope. Within the quantasomes are the green *chlorophyll* molecules—too small to be seen even with the strong power necessary to see the quantasomes. These are the molecules which trap the sunlight. The chlorophyll molecule is a very complex body consisting of atoms of carbon, hydrogen, nitrogen and oxygen, surrounding an atom of magnesium. The atoms are precisely positioned, any deviation in the pattern would nullify the function of the molecule. In orbit around this assembly of atoms is a cloud of negatively charged electrons. When the sunlight strikes a green plant leaf, or a chlorophyll molecule, one of the low-energy electrons may absorb a particle of light called a *proton*. This is the energy wave coming directly from the Sun. The electron, with this added energy becomes "excited" and moves so rapidly that it leaves the chlorophyll molecule. It is at this point that the photosynthetic process begins. Even though we do not know the details of the process, we know that it is extremely rapid in its movement for the excited electron which leaves the chlorophyll molecule remains in the excited state for less than 100 millionths of a second! But in this time, if we can call it that, it does its work and returns to join other electrons orbiting the chlorophyl molecule. There, it may again become excited as it and other electrons pick up photons. In a sense, the electron is simply a carrier of the light energy to the point where the next step in the process takes place. We must realize that there is a constant stream of electrons acting in this manner as long as sunlight is reaching the green leaf. It would be fair to state that this energy-in-motion is like a steady current of electricity.

"The next step of the photosynthetic process employs another conversion, the conversion of electrical into chemical energy. The energy released by the chloroplasts' electrons must now become a part of a chemical compound" (Pfeiffer, p. 36). The result of the process is the formation of *adenosine triphosphate*, or ATP. It is the "raw power" molecule which provides the energy for the cell to operate. We will find later that it is also the power molecule that provides energy for the animal cell. This very important molecule carries the energy or fuel that sparks the whole range of life processes, both plant and animal. It is found in all cells.

Possibly the most complex step in the photosynthetic process is the next one in which the ATP molecule goes into action, using water and carbon dioxide which the green leaf absorbs from its environment. In the chemical reactions that take place, the third phosphate group in the adenosine triphosphate molecule is released, providing a burst of energy. Other excited electrons from the chloroplast split water molecules into 2 H + O, providing free oxygen which goes into the atmosphere and hydrogen which combines with other atoms to form $TPNH_2$. Together, ATP molecules and the new $TPNH_2$ molecules react with carbon dioxide from the atmosphere to form the substance necessary to build and maintain plant tissues. The energy is provided by the ATP molecule as it breaks down. The end result of the process is the production of glucose, a form of sugar. "The glucose molecule contains a definite pattern of six atoms of carbon, 12 of hydrogen and six of oxygen—all of these constituents coming from the flow of raw materials absorbed by the cell from the atmosphere and the soil" (Pfeiffer, p. 37).

In this process there are many mysteries. One of them is the function of a protein molecule called an *enzyme*. There are a number of different kinds of enzymes, each seeming to have a specific function. Thus, only a certain enzyme can trigger the separation of the adenosine triphosphate molecule which in turn provides raw energy. Without this specific enzyme, manufactured by the ribosomes in the cell, there would be no chemical reaction to provide power for the growth of the plant and the production of glucose. This would stop the photosynthetic process and there would be no plant life, and most certainly no glucose to provide energy for animal life.

To say that the energy of the cell is locked up in the ATP molecule does not tell the whole story. When the specific enzyme trips the trigger to set off the breakdown of the adenosine triphosphate molecule, energy is released for use in the cell. We must realize, however, that the enzyme had to be "powered" to do its job and it had to be "directed" to the specific molecule "assigned" to it. Where did the first energizing power come from? How were the directions communicated? In what form did the energy flow into the cell for use by the nucleus and the other organelles which needed an energy supply to operate? To say that this was all under the direction and control of the nucleus is to evade the basic question, for the nucleus itself was in need of a power source as it is simply a combination of inanimate atoms.

The presence of the "Spirit-of-Life" provided by the Creator-God is demanded by the logic of the situation. When that spirit of life leaves the plant or animal cell, the character of the organic molecules becomes very clear. They are dead, they have no life. In time they decay and return to the simpler forms or even to their original atom form. The existence of life is locked up in the "sustained expressed will of the Creator-God". He said "Let there be." Apart from that expressed will, active in time, there would be no life.

We will see as we move ahead that all along in the life process there are points where without a proper enzyme at the proper point, or without a specific chemical reaction at a given point, life would be impossible. The scientific evolutionists appear to be all too ready to accept these as "coincidences," not as part of a design with specific goals.

"The existence of electrical energy-producing processes within the cell carries certain implications. Sustained electric currents cannot be obtained simply by throwing some parts together. Organization, plans and circuit diagrams are needed, just as they are in the production of man-made electricity. Nature makes use of precise plans and specialized equipment in the energy-producing processes in the cell; furthermore, powerful microscopes reveal details suggesting the existence of an even more sophisticated circuitry at the molecular level—the present frontier of cell research" (Pfeiffer, p. 35). What does this do to the case of the scientific evolutionist who cannot allow for design, plans or purpose in the coming-into-being of protons, electrons, elements, molecules, star systems, solar systems and life itself? In this whole story, it would appear to this observer, the one thing running like a red thread through everything, has been the evidence of design and the fulfillment of purpose, all in accordance with what we have called the sustained, expressed will of the Creator-God. It is the chloroplast with its chlorophyll molecules that holds the key to the existence and development of plant life through the photosynthetic process and this did not just happen to happen!

Now we must turn to the animal cell to learn how it is provided the power necessary for life. The structure of the animal cell is very similar to that of the plant cell—up to the point of the "powerhouse unit." In the animal cell, in place of the chloroplasts a different unit, the *mitochondria* comes into play. It, like the chloroplast in the plant cell, floats about in the cytoplasm of the cell. The food supply consists of the glucose which the animal ingests as it consumes plants—or other animals that have consumed plants. The mitochondria finds the glucose in the cytoplasm. The mitochondria transforms the glucose in a process which is in effect a reversal of the photosynthetic process. This action in the animal cell is called *respiration*. The animal "breathes" in oxygen and by the action of the mitochondria the glucose is burned and energy is released in the form of the same ATP molecule used in plant life. In the process, carbon dioxide and water are released. This completes what could be called a life cycle on earth. Plant life converts energy from the Sun into glucose which is stored in the fiber of the plant and as a waste product releases oxygen into the atmosphere. Animal cells must have oxygen to live and in using it and the glucose of the plant, obtain the energy they require for life, releasing carbon dioxide and water for reuse by the plant. Thus, as was stated before, plant life must have preceded animal life. In the long run, both types of cells need each other as each provides the essential ingredients the other needs for life.

Darting in and out of the spaces within the animal cell and near the structure of the ER are the mitochondria. There are as many as a thousand of them in each cell. They are in constant motion and often split apart, fusing end to end to form elongated bodies The function of the mitochondria is to generate the energy needed by the various parts of the animal cell, enabling them to function. "The mitochondria's raw material is the glucose absorbed by the cell and broken down in the cytoplasm. In a chemical operation, the cellular powerhouse "burns" the food to release energy and then loads this energy onto a chemical molecule with a jaw-breaker for a name: adenosine triphosphate, or ATP for short. This precious stuff—essentially raw power—is to cells what electricity is to the machines in a factory: it runs them" (Pfeiffer, p. 21). This ATP molecule is the same chemical compound which the photosynthetic process provided the plant cell. Thus both plant and animal cells receive their operating energy from the same chemical molecule.

There is much that is not known in the process of converting the glucose molecule formed in the plant by photosynthesis into the energy carrying ATP molecule for use in the animal cell. It begins in the semi-fluid cytoplasm of the cell where the glucose is broken down into more simple compounds. In a series of reactions the process of *glycolysis* occurs. The energy for the process is provided by ATP molecules from former operations assisted by a number of specific enzymes. Each molecule of glucose yields two molecules of *pyruvate*. In the process, two ATP molecules are used but in the new configuration of atoms after the production of the pyruvate, there are four molecules of ATP available for use. This means that once the process has been started it will produce the energy molecules required to continue.

The next step is known as the *Krebs Cycle* in which the mitochondria in the animal cell takes the pyruvate and transforms it in a series of reactions, possibly a dozen, into carbon dioxide and water plus thirty-six more ATP molecules! "This immense energy gain is accompanied by the production of carbon dioxide and water" (Pfeiffer, p. 39). The Krebs Cycle requires specific enzymes for each of the reactions in the process—which increases the puzzle about enzymes. It appears that the ribosomes under the direction of the RNA molecule form many kinds of protein molecules and produce some with specific functions. Just how this is done, how they are brought to the right spot at the right time is not known—but if they did not appear to trigger the chemical reaction that separates the adenosine triphosphate molecule and if the other specific enzymes did not appear at the appropriate points in the many steps involved in the Krebs Cycle which engineers the transformation of glucose into the ATP molecule needed by the animal cells for energy, there would be no life on Earth.

Huston Smith, professor of religion at Syracuse University, has an article, "Evolution and Evolutionism" in the July 7–14, 1982 issue of *The Christian Century*. He brings the mystery of the enzymes into sharp focus when he quotes Professor Wickramsinghe, of the Department of Applied Mathematics and Astronomy at Cardiff, Wales, in an article which he wrote for *Science News* (Vol. 121, January 16, 1982): "Some 2,000 or so enzymes are known to be crucial for life. . . . At a conservative estimate, say 15 sites per enzyme must be fixed to be filled by particular amino acids for proper biological function. . . . The probability of discovering this set by random shuffling is one in $10^{40,000}$, a number that exceeds by many powers of 10 the number of all atoms in the entire observable universe." Thus, in the opinion of this well known scientist, the probability of randomness in the functioning of the enzymes is simply out of the question—it could not happen. The scientific evolutionist may argue that the concept of "randomness" was abandoned with the development of the DNA molecule which directs the functioning of the cell and all of its parts. The argument is not valid, however, in view of the fact that no explanation has been given for the development of the DNA molecule other than that it is the result of random happenings to inanimate chemical molecules. Further, the insistance on the part of the scientific evolutionists on the concept that all changes resulting in the advancement of the various life forms occur because of the random mistakes or copying errors of the genes in transmitting instructions for reproduction, negates the argument that the randomness concept has been abandoned.

It is important to note that the plant and animal cells are almost identical—the difference being in the units that provide energy—and even here there are significant similarities. In capturing the energy of the Sun, it is the electrons which surround the chlorophyll molecule which become "excited" when they take on a photon, a sun-ray. In their excited state—when their energy load has been increased by the energy of the photon—they use the added energy in the making of the ATP molecule. In the animal it is again "excited" electrons that work in a sort of reverse action to convert the stored up energy in the glucose into the ATP molecule. "In the mitochondria, however, the electrons are originally excited by energy obtained from the Krebs Cycle (which in turn, gets the energy from pyruvate, the glucose product) rather than the energy in sunlight" (Pfeiffer, p. 39). Thus, both plant and animal cells receive the necessary energy molecules of ATP through the actions of electrons—the very tiny unit of energy that we first met in the Creation-Cloud after the Big Bang.

"Recent evidence from research centers suggests that the internal membranes of both mitochondria and chloroplasts are built in the form of mosaics of many thousands of pieces. Each basic unit of the mosaic is about two millionth of an inch in diameter and includes tadpole-shaped elements containing the individual acceptor molecules of the transport chain—Studies show that, in all cells, the molecules of electron-transport chains are placed close to one another in definite patterns to form assembly lines that yield ATP. This suggests that the controlled flow of charged particles in animal and plant cells, in living as well as man-made electronic systems, apparently requires a fixed arrangement of components and a specifically designed circuitry. —Flowing electrons, the tiny electronic currents which move along the electron-transport chains, are the essence of life in both plant and animal cells" (Pfeiffer, p. 40). Again, what does such a statement do to the theories of the scientific evolutionist, which insist that not only in "pre-life" matter but also in "life-development," progress occurs only by the mistakes or copying errors of the genes, which happen by pure chance?

In describing the significant parts or units within the cell—plant and animal—we promised to return to a more detailed discussion of the molecule of *deoxyribonucleic acid*, the DNA. It is found in all cells. It has been described as the "control center" or the "brain" of the cell. It directs the operation of the cell, the functions of the protein-making bodies and the reproduction process which creates new and identical cells. As 50,000,000 cells in the body of an adult human die every second and are replaced by an equal number of new cells, all in the proper places and at the right times, this function of the DNA is simply a matter of amazement. Remember that in all of this long description of the cells, we are dealing with an object (!) which is so small that it cannot be seen except with high powered microscopes! Yet, it is an exceptionally active, living body, with a great number of units, all of which work harmoniously together to produce the necessities for continuing life and reproduction. Only when the cell-formation process misfires, causing abnormal cells to develop, do we have a life-destroying situation, which we call cancer.

The nucleus of the cell is pocked with pores through which it sends out directions to the various bodies within the cell. It must also receive through its pores the substances it requires for its operation. Just how this is done is not exactly known but it has been determined that the nucleus governs the actions of the cell. The nucleus is filled with a mucous-like substance called *chromatin* in which is seen by way of great magnification, a thread-like network. These threads are called *chromosomes*. The number of chromosomes in plant and animal cells differs. For instance, in the human cell there are 46 chromosomes, in the cells of a mouse there are 40 while in a potato cell there are 48. Under very high magnification, the chromosomes appear to consist of strings of beads. These beads are called *genes*. There are about 1,250 genes on each human chromosome. It has been determined that the genes carry the specific information bits on heredity—they are sometimes called "sentences." In the process whereby a new cell is formed, the genes transmit the information they carry to the new genes of the new chromosome chain in the new cell. Thus, the parent cell is duplicated exactly. On very rare occasions, a gene may be damaged or misplaced, which would cause the new cell to be different, usually with dire results. This is called a mutation, a phenomenon which will be discussed later in this chapter. The network of chromosomes, if it could be drawn out and measured, varies in length depending usually on the sophistication of the cell organization. The human cell has a network length of three feet. The substance of the network or threads of genes is the deoxyribonucleic acid, or DNA.

How does it happen "by chance" that while all other human cells have 46 chromosomes, the male sperm cells and the female egg cells have only 23 each? We believe this to be another bit of evidence of design with a purpose—the creation of a new human being. Thus, when two human cells, male and female, unite, only the 23 chromosomes from each cell are involved in the make-up of the new cell—providing it with the required 46 chromosomes. At this point we might point out that the equality of the sexes begins at the very beginning! "Mathematically, there are more than eight million ways the 23 chromosomes of a human mother and the 23 of a father can combine. The odds that any two children of the mother and father will have the same complement of chromosomes is about one out of 70 trillion. And since each chromosome may have 1,250 genes, the odds against two identical individuals being born form such a large number that it has no name; it could be written out as 1 followed by 9,031 zeroes" (Pfeiffer, p. 58).

The DNA molecule is composed of just three ingredients. "There were simple sugars of a kind known as deoxyribose, there were phosphate units and finally there were nitrogen compounds. There were four of these compounds, or bases: adenine, thymine, cystosine and quanine, called A, T, C and G for short" (Pfeiffer, p. 60). It was not until 1953 when James D. Watson and Francis H. C. Crick published an article based on information from the studies of Maurice H. F. Williams, that an accurate description of the DNA molecule was provided.

"According to Crick and Watson, the DNA molecule is built something like a spiral staircase. The phosphates and sugars form the twisted frame of the stairs and the bases form the steps. Each step is made up of two bases joined in the middle. *But it is not a random pairing.* Adenine is always joined with thymine to form a step and guanine is always joined with cytosine. But the steps may follow each other endlessly and in any order, AT, GC, CG, AT, CG, TA, TA etc. A single gene might be a chunk of DNA stairway perhaps 2,000 steps long—and geneticists now think that it is the order of these steps, the arrangement of TA's, AT's, GC's, and CG's, that gives every gene its special character" (Pfeiffer, p. 60, emphasis mine).

When the time comes for a cell to reproduce itself, the DNA-ladder-like structure (to change the image from a twisted stair) separates down the center of the rungs, leaving each side of the frame with sets of half-rungs. Then each half-frame rebuilds a new second half identical to the one it lost in the split and each half-rung seeks out materials to rebuild the exact companion type it had before the split. That is, an A half-rung will seek a T half rung, a G half-rung will seek a C half-rung, etc. This is the only way they will rebuild and when all the rungs have been rebuilt there will be two DNA molecules which will be identical. The material for the formation of the new parts is found as *nucleotides* floating in the cytoplasm of the cell.

The astounding thing to think about is that in reproduction, not only is the DNA molecule divided into two identical units but that each finds itself as the basic material in a nucleus of a new cell which has all the component parts we have already discussed as the functioning parts of a cell—the membrane, the cytoplasm, the ribosomes, etc. In the plant cell there will be the functioning parts for the photosynthetic process—the chloroplasts with the grana and the chlorophyll molecules, while in the animal cell there are the functioning parts for the respiration process, the mitochondria, the Golgi Bodies, the Krebs Cycle mechanism, etc. The instructions for the division or replication of the new cell have all been transferred or built into the new cell!

The DNA molecule is carefully protected inside the nucleus of the cell. It "delegates its authority by producing a single-strand substance called *"messenger RNA"* . . . to carry the blueprints for protein construction to the ribosomes. DNA also makes smaller single-strand units called *transfer-RNA*, which are conveyers to bring the raw materials called amino acids to the assembly line where they are locked together in a proper sequence for building the particular protein ordered by the messenger-RNA. DNA's prodigious feat of producing precise directions in correct proportions at the moment they are required has been compared to the operation of a staggeringly complex computer" (Pfeiffer, p. 72).

To perform this delicate operation, a section of the DNA ladder unfastens itself—much like the separation that takes place when the molecule is reproducing itself. In this instance, however, just a portion of a "side" of the ladder separates out and in two lengths. Also, it is not exactly the same substance for it is *ribonucleic acid*, or RNA. The longer section is the carrier of instructions or the messenger unit while the shorter strand is the transfer unit which is, in a sense, the working agent. These two sections of RNA work together outside of the nucleus. The transfer unit seeks out the specific amino acids floating around in the cytoplasm of the cell and which have been ordered. It brings them to the messenger unit where they are attached in a precise manner in accordance with the code provided by the messenger unit. When the proper ingredients have been assembled, the ribosomes perform their task to make the protein molecule ordered. When it is formed, the protein molecule detaches from the RNA to perform its specialized task in the body. In the meantime, the DNA molecule has repaired its ladder, but at other points is repeating the act, producing a steady flow of RNA messenger and transport units to provide for the prodigious number of protein molecules needed for life. It could be said that it is at this point that the basic process of life-formation is taking place: the building of protein molecules.

"Given life, everything is possible." "Given life"—this is the crux of the argument! There is no logical explanation as to how life could have developed—except that it is "Given." The scientific evolutionist takes the position that "time is the hero of the plot"—given enough time, anything can happen. Given a billion years, the atoms in the universe, by random and accidental collisions, could form every conceivable size and shape of all possible molecules! These, in turn, would assemble themselves, by chance, into the extremely complex molecules of the carbon, hydrogen, nitrogen, phosphorus, oxygen and sulphur chains which would by magic or willy-nilly, form a life molecule which would be able to reproduce itself—meaning that it would be alive! Rather than accepting the scientific evolutionist's solution of the problem of the origin of life, I would suggest that it is not in the atoms of the DNA that the design of life lies, but in the will of the Creator-God.

The imagination of the various scientific writers appears to have run wild on the power of "slow and undirected" time being the answer to the problems they themselves have raised in connection with their hypothesis of scientific evolution. So often, having made a speculative statement, admitting that no proof was known, they have gone on to build a structure on the speculation as though it had been proven true. Thus, it becomes quite easy, given a billion years or so, to leap from the inanimate matter to a living cell—as Gribbin does (Genesis, p. 166).

In 1952, an American chemistry student, Stanley L. Miller, built in apparatus through which he circulated a mixture of water vapor, hydrogen, ammonia, and methane. He then had 60,000 volt electric sparks pass through this mixture. After condensing the resulting liquid in his flask, he found that it had reacted chemically to produce four amino acids which are called building blocks of proteins. This experiment excited the scientists because these compounds are necessary for the production of life molecules. But, *there was no life*! Dr. Robert Jastrow, in *Until the Sun Dies* (p. 62), said remorsefully that the one ingredient lacking was "time"—if only they could compress enough time and feed it into the flask, maybe a living cell might develop! Another thought should be added: this experiment was done under controlled conditions and an outside force, Stanley L. Miller, had a very important role in it. There were no accidental or random or willy-nilly collisions. They were contrived by man.

We promised earlier to take a careful look at the concept of "mutations" being the source of all evolutionary advances in life. In doing this, we would suggest that Charles Darwin should be studied more carefully. His work was not directed at the discovery of the origins of life, but rather on the development of life forms already in existence. The scientist in the last century has become a bit heady, what with all the very significant discoveries and advances that have been made in man's knowledge of the Laws of Nature, the physical laws that govern the Universe in general and our Planet in particular. In gaining this knowledge, it was rather easy to assume the agnostic position that only what can be proven in the laboratory can be accepted as true. This was the sign of the Newtonian Physics, the scientific method ruled. The scientist who is an agnostic—not all of them are—thus comes to what appears to be a logical conclusion, that all that is must be explicable by the mind of man. This leaves out of the equations of life any Transcendental Being who may have designed the Universe with a purpose and with goals. In effect, this position, as we have said earlier, reduces everything to the measure of the mind of man.

Darwin did a tremendous job in cataloging the enormous numbers of life forms which he found in his explorations around the Earth. His work, however, was one of description. He, and to a greater extent, his followers after him, made the mistake of confusing description with explanation. We have no quarrel with description—it is a necessary art and through it Darwin showed how the various species developed from the simple forms to the more complex. But—he did not explain the origin of life! To say that life developed in a "warm pond" in the primordial sea may be entirely true, but not without the "intervention" of the Creator-God!

And this Darwin allows! Nowhere have I read in Charles Darwin's *The Origin of Species*, or in *The Descent of Man*, that life, plant or animal, came into being spontaneously by the random collisions of particles of matter. I do find, however, that he believed that plant and animal life developed from one life form, *into which the Creator-God had "breathed" the essence of life.* In *The Origin of Species* he says in his concluding paragraphs: "Therefore, on the principle of natural selection with divergence of character, it does not seem incredible that, from such low and intermediate forms, both animals and plants may have developed; and, if we admit this, we must likewise admit that all the organic beings which have lived on this earth may be developed from one primordial form."

Moving from this point, we find in these same paragraphs the development of his thought with respect to creation and the development of life through the modifications of the original life forms, which is evolution. "I have now recapitulated the facts and considerations which have thoroughly convinced me that species have been modified, during a long course of descent. This has been effected chiefly through the natural selection of numerous successive slight variations; aided in an important manner by the inherited effects of the use and disuse of parts; and in an unimportant manner, that is, in relation to adaptive structures, whether past or present, *by the direct action of external conditions, and by variations which seem to us in our ignorance to arise spontaneously.... It appears that I formerly underrated the frequency and value of these later forms of variations, as leading to permanent modifications of structure independently of natural selection."*

This means that he did not claim that development came exclusively by natural selection but that modifications also came about by *"the direct action of external conditions"*—*"which seem to us in our ignorance to arise spontaneously."* Darwin here recognized that earlier on he had not given proper attention to these variations that came about apart from natural selection. He had *"underrated the frequency and value of these later forms of variation." They actually led to permanent modifications of of structure independently of natural selection!* This is of great importance. It is in this part of the evolutionary process where, as we have indicated earlier, the Creator-God and His workmen, intervened from time to time to give direction to the development of the Design for Life which He had laid out from the beginning.

It is interesting to note that Darwin complained about the reporting of the media in his day. "But as my conclusions have lately been much misrepresented, and it has been stated that I attribute the modification of species exclusively to natural selection, I may be permitted to remark that in the first edition of this work, and subsequently, I placed in a most conspicuous position—namely, at the close of the Introduction—the following words: "I am convinced that natural selection has been the main but *not the exclusive means of modifications.* This has been of no avail. Great is the power of steady misrepresentation." (My emphasis.)

Thus the steady misrepresentations of the views of Charles Darwin continue to this day. The most glaring example that I have seen is to be found in the opening quotes to the second chapter of Carl Sagan's *Cosmos*—where he purposely has distorted and changed the direct words of Darwin as found in the closing paragraphs of *The Origin of Species*. This is what Sagan tells us Darwin said: "Probably all the organic beings which have ever lived on this earth have descended from some one primordial form, *into which life was first breathed.*" Darwin indeed said the first part of this sentence, in essence, but not the second in this context. "There is grandeur in this view of life," Sagan continues, so far correct, but not in this place! Then he omits the following which Darwin said: "*with its several powers, having been originally breathed by the Creator into a few forms or into one.*" Sagan then picks up the balance of the sentence, "and that whilst this planet has gone cycling on according to the fixed law of gravity, from so simple a beginning endless forms most beautiful and most wonderful have been, and are being evolved." Sagan, to protect his scientific opinion that there is no Creator-God, deliberately altered the statement to leave out the word *Creator*.

This leads us to the final paragraph in *The Origin of Species*: "There is grandeur in this view of life, with its several powers, *having been originally breathed by the Creator* into a few forms or into one." Darwin thus believed that the Creator-God breathed life into the first life form. The Creator-God has followed up by providing everything necessary to insure that life would continue to develop in all of its many forms, intervening only at those points wherein "direct action" was required to effect specific changes.

Finally, in *The Descent of Man*, we read: "I am aware that the conclusions arrived at in this work will be denounced by some as highly irreligious; but he who denounces them is bound to shew why it is more irreligious to explain the origin of man as a distinct species by descent from a lower form, through the laws of variation and natural selection, than to explain the birth of the individual through the laws of ordinary reproduction. *The birth both of the species and of the individual are equally parts of that grand sequence of events, which our minds refuse to accept as the result of blind chance.*"

Thus, Charles Darwin does not stand with the scientific evolutionist who today stoutly affirms that all development of life forms has happened "willy-nilly" as stated by Gribbin. The intervention of the Creator-God at significant points in life development by direct action, not by blind chance, has been consistent from the beginning until now.

Huston Smith makes some very interesting remarks in "Evolution and Evolutionism." "The irony is that the evolutionists have no plausible theory to pit against religious accounts of human origins. Their discoveries show a history of evolutionary advance but do not explain how or why that advance occurred." As we have stated a number of times, the best they have come up with for the origin of life is that "somehow, somewhere" the first self-replicating molecule appeared. Adding to this weak position they have made it abundantly clear that they believe that no improvements in the life forms have occurred except by the accident of genetic mistakes!

In the discussion on life development, Gribbin speaks of the "invention" of sexual reproduction as being of great significance in producing the diversity we find in life today. Then, almost as if in haste, he adds the footnote: "Remember that "invention" is used as shorthand for "a succession of accidental favorable mutations (copying errors) which produced organisms more efficient at surviving and reproducing than their competitors. *No conscious planning is involved*" (*Genesis*, p. 219). His position is clear: every upward step in the evolutionary process is the result of a copying error in the genes of the DNA molecule. There is no place, in the scientific evolutionist's theory, for the influence of a Creator-God nor the direct action by his guiding hand or exercise of his will.

"Mutations," we are told by Gribbin, "Do not happen suddenly, producing dramatic physical changes in the body of a new individual, compared with the bodies of its parents. Nor do mutations happen "in response to" environmental changes" (*Genesis*, p. 22). Thus, mutations bring about changes in a given body by the method of the slow accretion of genetic mistakes. But, nagging questions persist. Take as an example, the invention of sexual reproduction that was just referred to. The genetic mistake process must indeed have been long and complex. New sets of physical organs which, to begin with, would have had no meaning or value would have to be developed to provide for the male sperm to be produced. Ways and means would have to be provided for the sperm to travel from its place of origin to the point where a female egg would be located for the purpose of fertilization. Nerves, muscles, tissues, physical structures would all have to be formed—long before there would be any use for them! How such a process makes a body "More efficient at surviving" escapes me! And this is only half of the problem for if such a system were to be effective, there would have to be a proto-female body which would have to be experiencing similar kinds of genetic mistakes that would provide for a much more complex set or organs, blood vessels, nerves, etc. Eggs for fertilization would have to be produced on a regular basis. A place—womb, in the mammal, would have to be prepared to receive the fertilized egg and to nurture it in its gestation period. In an egg-laying animal, the process would have to provide the mechanism for laying eggs and a system for hatching them!

The problems involved are almost innumerable—the period of gestation would have to be determined, and the mechanisms in the birth process would have to be determined, and the mechanisms in the birth process would have to be worked out. But—almost more problematic—how would this genetic error process, that would develop slowly over a long period of time, be able to guarantee that a female body would be in place at the right time to receive the sperm from the male? If all of the above—and there would actually be much more—were to have happened by "chance mutations," the laws of probability would have simply broken down along the way. The concept that chance, working alone, could accomplish the invention of sexual reproduction is not feasible. A process leading to a specific goal implies a design with purpose. Such a process does not develop by happenstance nor can it be achieved by the random mistakes in genetic functions. The scientist may describe the process but description is not explanation. There was Intelligence with a purpose and a goal that designed and activated the development of reproductive systems.

Smith cites another kind of situation when he raises the probability difficulties that would be involved in the development of birds from reptiles all on their own, willy-nilly. "The number of generations through which a large number of immediately disadvantageous variations would have had to persist in order to turn reptiles into birds, say scales into feathers, solid bones into hollow tubes, the dispersion of air sacs to various parts of the body, the development of shoulder muscles and bones to athletic proportions, to say nothing of conversion to a totally different biochemistry of elimination and changeover from coldblooded to warm—makes the notion of chance working alone preposterous. As Professor Pierre Grasse, who for 30 years held the chair in evolution at the Sorbonne, has written: The probability of dust carried by the wind reproducing Durer's "Melancholia" is less infinitesimal than the probability of copy errors in the DNA molecule leading to the formation of the eye; besides, these errors had no relationship whatsoever with the function that the eye would have to perform or was starting to perform. There is no law against day-dreaming but science must not indulge in it" (*Evolution of Living Organisms*, Academic Press, 1977.

Finally, on the subject of the development of advanced life forms by chance mutations we should point out that as these are all supposed to be accidents to the genes or errors by the genes in the transmitting of information, it follows that there can be no relationship between the accidental errors in any manner whatsoever. Thus in a mutation that involves, say the development of the bird from the reptile, there would have had to have been great numbers of unconnected errors which somehow would act on a "started mutation" in a specific reptile family group to carry through the mutation to the completion of the bird. To say that the mutation was not a chance happening but one which came within the scope of the sustained, expressed will of the Creator-God would resolve the problem. It would also be in accord with Charles Darwin's statements as quoted earlier.

We have come to the point in our study when we can say that life has been established. Under the guiding hand of the Creator-God and in the fulfillment of his purpose, the development of life in all of its varieties and complexities was assured. We need not follow the process further—the results are evident all about us.

An interesting new concept put out by scientific evolutionists in an attempt to get themselves out of a difficult spot of their own making is the Gaia Hypothesis. It appears quite evident that at least some of the scientific evolutionists we have used as references are not really satisfied with having "pushed god off the edge of infinity," as Davies put it. To a non-scientific observer, the evidence supporting the concept of a Creator-God being responsible for the existence of the Universe as we know it is far stronger than the evidence which the scientists claim for the support of the concept that the Universe just "happened to happen" or that it came into being totally by chance.

The several defensive statements we have quoted in which the public is warned against anthropomorphic thinking, emphasizing that human life developed through a series of "bizarre mistakes," indicate that the scientists themselves are struggling to put down the evidence they have developed which indicates the presence of an Intelligence, a Creator-God. Nowhere is this ambivalence more clearly shown than in the statement made in Gribbin's last chapter on the Gaia hypothesis. He refers to Jim Lovelock, the British scientist who has developed this concept which attempts to explain "the ecological balance between the different kinds of living organisms on Earth" in which "over thousands of millions of year, life on Earth has helped to maintain a stable situation with conditions well suited for life. . . . Could such stability really have been simply good luck? Lovelock argues that this is unlikely, and points to the aweful examples of Venus and Mars as planets left to their own devices, with no living regulators to maintain this kind of stability" (pp. 305f).

This is a remarkable statement. The concept, as well as this reader can understand it, appears to propose that when the first so-called self-replicating molecules "somehow, somewhere" appeared on Earth, a kind of "spirit-of-life" developed, which over all the thousands of millions of years since has in an "ebb and flow" type of movement, guided the development and maintainance of not only life on the planet but in some manner all the "natural rhythms"—like the Milankovitch Cycle of recurring ice ages. That the Gaia figure is indeed a substitute Creator-God for the scientific evolutionist is very clearly indicated in the following passage of Gribbin's *Genesis*: "There is still another 5000 million years or so before our Sun comes to the end of its main-sequence life and expands as a red giant to engulf the inner planets, and there is no reason to expect the extinction of life on Earth—*the death of Gaia*—before that time. There is, though, no reason to expect human life to survive for even a fraction of that time. We may be victims of a natural disaster—a nearby supernova explosion, perhaps, producing a flood of radiation and massive faunal extinctions across the Earth—or of our own folly. *If so, given the events of the past 65 million years, Gaia will have ample time to try again to produce an intelligent 'nervous system' from some other species*. We may be just the first of many intelligent forms of life to develop here" (pp. 311f, emphasis mine).

This is a sophisticated description of a "God-figure"—one that could "pick up the pieces" after a nuclear war had destroyed life as we know it and "set out again to produce an intelligent nervous system." Gaia is the Greek word meaning "Earth Goddess." To carry this point one step farther, Gribbin introduces the suggestion that it is in our own interest to be on the right side of Gaia: "It is for our own sake, according to the Gaia hypothesis, that we should work with, rather then against, the natural systems which maintain the comfortable environment on our planet" (p. 306). We would add, it is always best to be on the side of God!

All of which means that the scientist, too, needs to have God to account for the Earth and life on it. By implication, the hypothesis would have to extend also to the whole Universe. Not only must the scientist have God in his equations but there is a need, inborn in the being of a human—and I strongly suspect in the plants and animals as well—to have a God to bring stability to all we know. The day appears to be at hand when the scientific and the religious may grasp each others hand and walk bravely into a future that has a purpose and goals that now are known only by the Creator-God.

CHAPTER XII
THE GREAT INTERVENTION

Life development after the cell became established, in accord with the sustained, expressed will of the Creator God, no doubt followed the course as outlined by the scientific evolutionists. Each step along the way to the development of life as we know it today was made in fulfillment of the design and purpose set in motion by the Creator-God when the Big Bang occurred. To one living in Universe time, it appears that the process has been very long, the progress very slow. We must realize, however, that for the Creator-God, this is not so. For one who lives in eternal time, no time at all has passed! The concept boggles the mind of one whose understanding of time has been limited to Universe time, just as the concept of eternity—a never-ending-never-beginning process from everlasting to everlasting—is beyond the comprehension of one who has only lived in Universe time. Universe Time began with the Big Bang and will end when the Creator-God no longer sustains it. To measure the period of its existence relative to Eternal Time would be without meaning.

It would appear to me that the scientific community has not been responsible in facing the issue of the ultimate cause of creation and of the development of the Universe after the Big Bang. Scientists appear to be saying that they cannot believe in an Eternal Being with no beginning or ending because it is incomprehensible that such a Being should exist. At the same time, they ask people to believe that this Universe simply exploded into being out of nothing and without a First Cause and that it developed from that point to the present without any relationship to a Supreme Being and that eventually it will return to nothing! They simply do not face up to the issue. In their efforts to avoid the issue they have adopted a policy of obfuscation—pushing all evidence of design with purpose out of view, obscured by the fog of nameless time. Their position has been dogmatic.

And yet, their ambivalence shows from time to time, as we have pointed out earlier. One scientist, Richard Morris, in *The End of the World* (p. 133), hedges the scientific position: "Since discussions of the big bang theory often use such terms as 'creation' and 'the origin of the universe,' it can sometimes seem that we are no longer talking about physics, but are trespassing into the field of theology. Therefore, a word of caution is necessary. When scientists speak of the 'origin of the universe,' they refer to the beginning of the universe as we know it. *They do not mean to imply that nothing existed before a certain point in time—only that they know of no way to extrapolate backward beyond that point.*" This is a good statement, but many scientists don't put it that way and specifically state that there was no First Cause and that the Universe started of itself and will end by itself. Even Morris goes on in the next paragraph: "To ask what happened before the big bang is to pose a question that cannot be answered. There may not even be anything as 'before'; time itself may have come into existence in the initial explosion." His ambivalence is apparent.

The Church leadership has also exhibited an unwillingness to face demonstrable facts and has been equally dogmatic in the attempt to prove or hold to a position. Dr. Philip Heffner in *The Lutheran* (April 20, 1983) suggested that theologians and scientists should try to share their knowledge because "There's a horrible lack of understanding in each group about what the other believes and does."

The Creator-God having expressed his will in the Big Bang of creation, that there be a Universe such as we have, has sustained his expressed will to this day. There is no reason to believe that he will not continue to sustain what he has willed until his purpose has been fulfilled. It would appear that the chief purpose of the Creator-God was to develop mankind in his own image and likeness. It could be that in Eternal Time mankind has a role to play!

Having put into place the "Laws of Nature" at the beginning of Universe Time, the Creator-God was bound by his own will to allow and direct the development of the Universe as we know it. This means that in an orderly manner—even though there was much violence in the process—the Creation-Cloud provided the energy forms (protons, electrons and neutrons) which would form the elements of matter. In addition, it provided the photons for the production of light and the neutrinos whose functions are not yet clearly understood. Out of these primary energy forms all of the Universe, including our Galaxy, our Solar System, our Sun and our Planet Earth have been developed through Universe time. This development can be called evolution, the building up from simple forms to the more complex. This was done in the inorganic, pre-life period continued in the organic period on to the present date. Evolution has not stopped and will not as long as the Creator-God continues to will it. This building of the Universe as we know it was done in an orderly manner and under the supervision of the "workmen" of the Creator-God, as we have indicated. The whole subject of angels, their existence and functions must be brought out of the darkness of the medieval age and studied in the light of today.

In the period of time preceding the advent of life, evolution involved the development of matter, the physical stuff of the Universe. Out of this matter came all the rest—the Universe as we see it, including life as we know it on this planet. We do not know that life exists at any other point in the Universe. If the scientific evolutionist's theory of development of life by happenstance were correct, there would be no logical reason to preclude the development of life in the far reaches of our Galaxy and in the ever more remote spaces of the billions of other Galaxies. On the other hand, if the Creator-God is responsible for this Universe, it just might be that only one planet in all of the Universe has life. This point, however, should not be pressed, for if it were in the will of the Creator-God to develop other life-bearing planets—so be it!

The fact that the evidence, as we have seen it, indicates that the Creator-God is responsible for the Universe, and that he established from the beginning the Laws of Nature which brought order out of chaos, and that he guided the development at all times, does not preclude the possibility that he might enter into the process directly at any time. In fact, it would be most illogical not to expect that the Creator-God would do this, for if it were his will to establish a Universe in which life as we know it were to exist, the reasonable attitude of the Creator-God would call for him to give special guidance to the process from time to time—to intervene in it, to give impetus to necessary changes during the course of evolutionary development.

Many possible points of intervention could be suggested, such as the triggering of the Big Bang explosion of Creation, the breakup of the Creation-Cloud to form the Galactic System, the triggering of the separation of our Solar System cloud from the spiral arm of the Milky Way Galaxy, the formation of the Planet Earth with all 103 natural elements, the giving to inanimate atoms the capability to form a "life molecule," the provision of the photosynthetic process within the cell for the purpose of providing energy from the Sun for the first life form, the provision of the cell as the body within which the life molecule would live, the "breathing of life" into the first life form when the time was right, the implantation of the blueprints of life in the DNA or life molecule, or to (leap over a long period of time) *the intervention of Himself into the life of mankind in the Person of Jesus Christ.* This event, which usually has been dismissed by the scientific community as religious superstition, needs to be looked at scientifically. The use of the scientific method in the examination of the evidence should be encouraged. The act of the Incarnation, wherein the Creator-God in the Second Person of the God-head was born of a woman, changed the whole relationship between the Creator-God and the human race. For through it the Second Person, whom we know as Jesus Christ, became true man even as he is true God! We do not yet know the full significance of this act, but one thing becomes quite apparent: that the human is destined to share Eternity with the Creator-God, his workmen and whoever else might be in the Eternal Design. The question of life on other planets in our Galaxy or any Galaxy becomes a bit doubtful in the light of the "intervention" of the Creator-God into the life of mankind on our planet. It is doubtful because to do that more than once does not appear to be a logical course. Nevertheless, while doubtful, it need not be ruled out.

The "new" or different relationship that was to exist between the Creator-God and man with respect to life after death was reflected when, on the cross, Jesus said to the penitent thief who was also crucified that day: "Today thou shalt be with me in Paradise." These words plainly mean that, while death would separate them from their bodies, they would meet that very day as persons in a setting in which they would know each other! Life in this Universe may end, but life in the Eternal setting continues.

Many may ask why Jesus, as the Second Person of the Godhead, could not have escaped death on the cross. He could have—but that would have defeated the whole purpose of the Creator-God in creating the Universe and in placing in it the human who had been made in his image. This is made very clear in the scene in the Garden of Gethsemane when Jesus was arrested to be taken to trial before the Sanhedrim. One of the disciples drew his sword and struck a servant of the high priest. We read in Matthew (26:52–54): "Then said Jesus unto him, Put up again thy sword into his place, for all they that take the sword shall perish with the sword. *Thinkest thou that I cannot now pray to my Father, and he shall presently give me more than twelve legions of angels? But, how then shall the scriptures be fulfilled that thus it must be?*" An army of angels was at his command, but to call them into action would destroy the plan of the Creator-God for the human race and for the whole creation. While there is much more that we do not understand, some things are clear. It was in the "game plan" for the Christ to go through the torturous death on the cross, to die and to be raised again. Jesus had power over life and death. He said: "Therefore doth my Father love me, because I lay down my life, that I might take it again. No man taketh it from me, but I lay it down of myself. I have power to lay it down, and I have power to take it again. This commandment have I received of my Father" (John 10:17–18).

Tied in with this "game plan" there is the whole matter of the existence of *evil* in the Universe. While I cannot go into that in this context, we know that evil existed at the time of the birth of the human race. It may have existed long before, throughout Universe time and even back into Eternal time. The many references to the Devil and his angels in both the Old and New Testaments give evidence that the struggle between God and Satan has been on-going. The apparent contract or agreement between God and Satan relative to the testing of Job is very significant. In the life story of Jesus Christ, the temptations of the Devil seeking to derail the course of the Great Intervention, are of great significance, coming from one who almost appears to know what it is like to be God!

The question of the existence of evil in this created Universe is one that cannot be precisely answered by someone who lives in Universe time. But a guess can be made. Lucifer, the fallen archangel, who is the Devil, was a chief among the angels. When the Creator-God made known to the angelic hosts his decision to make Man in his own image, Lucifer became jealous and led a revolt among the angels. With his followers he was expelled from the fellowship of Heaven, but not destroyed. He still exists and is still revolting, seeking to destroy the Man who was created in the image of God.

The fact that Lucifer could rebel indicates that the potential for evil has always existed. This is the risk factor that the Creator-God had to allow. To have excluded the potential-for-evil factor would have destroyed the validity of the potential-for-good. It is the will of the Creator-God that all of the Universe be in harmony with him, not by the force of coercion but by an act of free will. Why the Creator-God did not destroy Lucifer and his followers at the time of their rebellion can only be understood in light of what we have learned of the love of God as it has been revealed through Jesus Christ, "who came into the world that the world should be saved through him." In the unfolding of the story of the human race and the Intervention of the Creator-God into the human life-line, we are given the word that the Creator-God has defeated Satan and that in Universe time that defeat will become final.

The evidence is overwhelming that Jesus Christ was not just another man, a great teacher, a philosopher with great wisdom. *He was another kind of Being—an out-of this-world Person who had come into this world for a purpose.* This gives meaning to the passage in the first chapter of Genesis where the Creator-God says: " 'Let us make man in our image, after our likeness.' So God created he him; male and female created he them." As we indicated in Chapter Nine, simple nomadic people thousands of years before the birth of Jesus Christ could possess the knowledge revealed in the first chapter of Genesis only if they were "given" the message.

It is necessary for us to look at the evidence which demonstrates that Jesus was not just a superman but rather a Creator-God-Man. This should be done by brushing aside the religious connotations and looking at the evidence objectively, supported as it is by many contemporary witnesses. We should remember that the people who were followers of Jesus Christ while he lived, and remained so after his death and resurrection, had nothing to gain, materially speaking, by their allegience to him. Instead they faced the loss of their material possessions, physical suffering and often death. They were not a fanatical group but rather conservative and, to begin with at least, a bit scared!

The first four books of the New Testament in general present running accounts of the life and teachings of Jesus. They do not agree in all details; some contain accounts which others leave out. This is what we should expect when four people write about the same subject, especially when we look at the differences between them and consider that two of them may never have seen Jesus. All of the writers in these books describe events which occurred after the resurrection. His friends were skeptical when they first learned of it. In fact, he had to prove to them that he was "for real" when he appeared to them in a group! We will call attention to a few of these incidents.

John wrote of the disciple Thomas, who had not been present when Jesus first appeared to a number of other disciples after the Resurrection. When told of the event, he said, in effect, "No way will I believe that Jesus is alive, unless I can touch the nail-prints in his hands and place my hand on the wound in his side." Note that these disciples were scared. They met in a room with barred doors; they were realistic about the danger they faced at the hands of the authorities who had put Jesus to death. Eight days after Thomas had expressed his doubt, Jesus suddenly appeared and spoke directly to him, telling him to touch his hands and his side. All in the room were witnesses to the event; it was not a fantasy or confabulation. The body of Jesus was real—yet he had entered a locked room. His person was real—he knew what Thomas had said a week earlier. He spoke to Thomas and the others, yet his body was not subject to the stone and mortar of the building; he was able to appear and disappear at will. John tells us in this same account that Jesus gave many other signs in the presence of the disciples and others—too many to record, but that some had been recorded so that people could know that Jesus was the Christ, the Son of God.

Luke, the physician—and thus a man of science in his day—recorded the following incidents. On the day of the Resurrection a group of women went to the tomb to anoint the body of Jesus. They fully expected to find his body in the tomb, but they found it empty, the enclosure stone rolled away. As they stood there, two angels "in shining garments" appeared and said, "Why seek ye the living among the dead? He is not here but is risen." The women hurried into the city to tell the disciples, who were evidently staying together in a body, fearful that they would also be put to death. Luke says, "And their words seemed to them as idle tales and they believed them not." Nevertheless, Peter ran to the sepulchre and saw the linen clothes that had been wrapped about the body of Jesus "laid by themselves" as though they had been folded and placed on the bench of the tomb. John, in his account, says that the napkin that had been tied around the head of Jesus was not with the other clothes but was "wrapped together in a place by itself." The accounts, a bit different in small details, appear to be authentic; there were a number of people who witnessed all that happened.

Luke also tells of the two men who, on the afternoon of Resurrection Day, walked to the village of Emmaus, a short distance from Jerusalem. As they talked of the strange things that were happening, a third man joined them. They did not recognize him as Jesus. When he asked what it was they were talking about, they told him of the crucifixion of Jesus and of the hope they had had that he should have been the redeemer of Israel. They also told him of the strange news they had heard that morning—that angels had appeared and had told a group of women at the tomb that Jesus was alive. Then, this stranger, scolded them for being so slow to believe. He went on to tell them what the scriptures had foretold about the Christ. When they arrived at Emmaus, the two asked the stranger to stay with them. As they sat at a table to eat the evening meal, he took the bread and blessed it. Then they realized who he was and as they did, he simply vanished! They left immediately to go back to Jerusalem to tell the disciples that they had seen Jesus alive. Again, the story has the ring of truth. Again Jesus demonstrated that time and space had no hold on him.

Returning to Luke's account, after the men who had gone to Emmaus had told their story, "Jesus himself stood in the midst of them, and saith unto them, Peace be unto you. But they were terrified and affrighted, and supposed that they had seen a spirit. And he said unto them, Why are ye troubled? and why do thoughts arise in your hearts? Behold my hands and my feet, that it is I myself. Handle me, and see; for a spirit hath no flesh and bones, as ye see me have. And when he had thus spoken, he showed them his hands and his feet. And while they yet believed not for joy and wonder, he said to them, Have ye here any meat? And they gave him a piece of broiled fish and of a honeycomb. And he took it and did eat before them" (Luke 24).

Thus, Jesus had to prove to his disciple that he was truly alive, that his body was real. And yet we see from these incidents that while he was real he also had an unreal quality—he could appear and disappear at will, barred doors did not prevent his entrance into their room. Also, he knew what they talked about when he was not present. He was, as we said earlier, another kind of Being, an 'out-of-this-world" Person.

And it was not only in the post-Resurrection period that he had been like this. All through his life he had been doing strange things, which somehow were not really recognized as such. Again, we will cite some incidents that indicate that while he was truly a person, he was much more. Luke, in chapter 7, tells of the Roman centurion whose servant was sick unto death. The elders of the Jews asked Jesus to heal the man, for the centurion was their friend. Jesus went with them on the way to the centurion's home. "And when he was now not far from the house, the centurion sent friends to him, saying unto him, Lord, trouble not thyself: for I am not worthy that thou shouldst enter under my roof. Wherefore neither thought I myself worthy to comeunto thee; but say in a word, and my servant shall be healed. For I also am a man set under authority, having under me soldiers, and I say unto one, Go, and he goeth; and to another, Come, and he cometh; and to my servant, Do this and he doeth it. When Jesus heard these things, he marvelled at him and turned about and said unto the people that followed him, I say unto you, I have not found so great faith, no, not in Israel. And they that were sent, returning to the house, found the servant whole that had been sick." The mind of Jesus had sent an unspoken message to an unseen person, and instantly the sick servant was well. Here Jesus demonstrated not only his power to heal diseases, but that his power moved with the speed of thought. He willed, and it was done!

Again, in the same chapter, Luke continues: "And it came to pass the day after, that he went into a city called Nain; and many of his disciples went with him and much people. Now when he came nigh to the gate of the city, behold, there was a dead man carried out, the only son of his mother, and she was a widow; and much people of the city was with her. And when the Lord saw her, he had compassion on her and said to her, Weep not. And he came and touched the bier; and they that bore him stood still. And he said, Young man, I say unto thee, Arise. And he that was dead sat up and began to speak. And he delivered him to his mother. And there came a fear on all and they glorified God, saying, that a great prophet is risen up among us, and, that God hath visited his people."

The power that Jesus possessed goes far beyond anything that science can explain, then or now. Rather than dismiss it as fantasy, however, scientific study should be made to determine the accuracy of the reporting. Jesus worked in the open; there were witnesses to what he said and did. Years later, when Paul was on his way to be tried before Caesar in Rome, he spoke before King Agrippa. In testifying to the life, death, and resurrection of Jesus, he said, "These things were not done in a corner."

The Creator-God, in Jesus Christ, had intervened in the life of mankind. This was in keeping with the plan set in motion at the time of the Big Bang and affirmed in a number of ways, specifically in the Genesis passage quoted earlier. The relationship between the Old Testament and the New is remarkable. The writers of the Old Testament lived in isolation from each other for the most part and there are centuries of time between the various book. And yet they all in one manner or another point forward to the birth or coming of Jesus—the intervention of the

Creator-God into the human family line. The New Testament records the fulfillment of the Old Testament promises, telling the story of the incarnation and birth of Jesus Christ. The New Testament reveals to mankind the "Personhood" of the Creator-God. Jesus also very explicitly identified his relationship with the Creator-God, his Father, in many ways, but none more specifically than when he spoke to them as recorded by John in the fourteenth chapter: "Let not your hearts be troubled: ye believe in God, believe also in me. In my Father's house are many mansions" if it were not so, I would have told you. I go to prepare a place for you. And if I go and prepare a place for you, I will come again, and receive you unto myself; that where I am, there you may be also ... I am the way, the truth, and the life: no man cometh unto the Father but by me. If ye had known me, ye should have known my Father also: and from henceforth ye know him, and have seen him ... Have I been so long a time with you and yet thou hast not known me, Philip? He that hath seen me hath seen the Father; and how sayest thou then, Show us the Father? Believest thou not that I am in the Father, and the Father in me? The words that I speak unto you, I speak not of myself: but the Father that dwelleth in me, he doeth the works. Believe me that I am in the Father, and the Father in me; or else believe me for the very works' sake."

The primary purpose of the Creator-God in the creation was to bring into being the human race, which was to have a physical relationship with the spiritual reality of the Creator-God. The result of this union or intervention was to translate the finite-bound human race into the infinite capability of the eternal Creator-God. Once Jesus Christ had been born of a woman whose human egg had been made alive by the power of the Holy Spirit—the Third Person of the Godhead—humankind would never again be the same. The Creator-God had identified with the Human—taking on the physical body of the humankind he had created. Universe time will end, but the human race will have been prepared for life in eternal time.

We find the theological message of the Old Testament in the narrative description of the development of humankind from the beginning of the race to the time of the birth of Jesus. At no time does this narrative reveal a "primitive man." From the beginning, the story of mankind in general is told, indicating the three major racial types, the Caucasian, Oriental and Negroid. With Abraham, however, the story-line shifts to the description of the development or evolution, of a specific family line which we know as the Hebrew people. It was into this family, specifically the tribe of Judah, that the Creator-God was to intervene. This was first foretold in Genesis 49:10. From that time on, the Messianic line was traced through King David down to Mary who was of the family of David, of the tribe of Judah. The "Intervention" occurred when Mary became pregnant by the Holy Spirit. In this way, the human race was inducted into the life-line of the Creator-God and thus into an eternal relationship. All this was a part of the sustained expressed will of the Creator-God.

The astonishing truth is that God exists! He is like Jesus Christ, who said, "He who has seen me has seen the Father; I and the Father are One." He is responsible for the creation and preservation of the Universe. His sustained expressed will is the enabling power of the Universe, administered through the hosts of angels who do his will in accord with the will of God and the Laws of Nature which he has established.

So, as you stand on your high point, looking up at the stars, do not be afraid. Rather, be glad! The Creator-God has brought this marvelous Universe into being and has sustained it until this day. It will come to an end when his purpose has been fulfilled. In the meantime, he continues to work in it, together with his workmen, the angels. We do not know what wonders may yet be revealed. Our human race has been joined by flesh and blood to the life of the Creator-God! As members of his Universe-time family, we will live forever in Eternal time, a dimension that we cannot now comprehend.

Peter—a man who was unschooled, a fisherman before he met Jesus, a man who knew nothing about atomic reactions or nuclear fission, wrote in a letter to the churches: "But the day of the Lord will come as a thief in the night; in the which the heavens shall pass away with a great noise, and the elements shall melt with fervent heat, the earth also and the works that are therein shall be burned up . . . Nevertheless, we according to his promise, look for a new heavens and a new earth wherein dwelleth righteousness" (2 Peter 3:10–13).

POSTSCRIPT
THOUGHTS ON QUANTUM MECHANICS

Science and religion must come up with answers to the questions as to the source of the energy of the Big Bang and the manner wherein this energy developed into matter and subsequently into the Universe as we know it. The theories of quantum mechanics appear to have led the physicists a long way in the right direction. As we will see a bit later, the theorists are about at the end of the road, without quite having achieved their goal.

Quantum mechanics is a vehicle that enables us to "see" how the existential will of the Creator-God, working-pulling-attracting, provides the energy that is responsible for the development of this Universe of matter and motion. We call that force the power of gravity. We have given it a non-scientific description as a measure of the level of attraction between the Creator-God and the mass of the Universe. We say that the sustained expressed will of the Creator-God translates into a medium that is comprehensible in the Universe as we know it—wherein the human, in the image of the Creator-God, is able to come to understand the "how" of the evolution that brought the Universe into being. The image of the particle-wave hovering between "being" and "not-being" helps us to "see" how matter is produced. Quantum mechanics leads us to understand how, through the particle-wave phenomenon, the transformation of the energy expressed in the Will of the Creator-God takes place and can be "known" by the mind of man.

As we read the intriguing account of particle interactions, though we don't understand what is going on, we get the intuitive feeling that here we are at the point wherein the expressed will of the Creator-God is being quantified as particles of energy which will assemble as a proton, carrying a positive charge, and an electron, carrying a negative charge. The proton also carries the mass of the "oncoming" atom of matter, hydrogen. The destiny of the proton is to be captured by the electron. Thus, the two are "married"—made one in the hydrogen atom, the first of the elements of matter.

We also have the feeling that this process is identified in the equation $E = MC^2$, where the negative electron plus the positive proton in a relationship with Time (the speed of light squared) forms the first atom of matter as we know it. We think also of our setting of the equation in theological terms: the sustained expressed will of the Creator-God = the Spirit of God × active in Time ($E = MC^2$). Out of this we get the picture of the evanescent transition of the will of the Creator-God moving through the particle interactions which appear to be a hovering of particle-waves (energy packets), from which there emerges the proton, ready for the fulfillment of its destiny, and the electron, seeking its mate. All of this happens so rapidly that we observe it as instantaneous action. The Creator-God willed it and it was so!

The Newtonian epistemological system, wherein deterministic laws were set out, concerned itself with the discovery of the laws of nature that described the matter of the Universe in specific terms. The scientific method was the trademark of the system. Equations had to balance—cause and effect ruled the Universe. The system worked well as long as the Universe was looked at in a macrocosmic manner. Troubles began to surface as scientists realized that the atom was not the fundamental unit but could be broken down into an electron and a proton—so small that no one has seen either one! Such entities (if they should be called that) could not be handled in the same manner as the atoms of the 103 elements of the Universe. The scientific method was not applicable. Along with this discovery had come an invaluable tool, the electronic computer, which enabled the rapid calculation of vast quantities of whatever was quantified. In no way could Newton's laws apply to the submicroscopic universe that was rapidly opening up to the scientific world.

Quantum mechanics is concerned with the description of the submicrocosmic realm of the atoms and their components. It should be said that this system could not have been developed without the aid of the electronic computer—something to think about. Also, it should be noted that this theoretical system in no way discredited the Newtonian physics—it simply picked up the descriptive task of classical physics at the point where it had come to an end. It is interesting to note further that before the deterministic laws of Newton became meaningless as applied to the microcosmic universe, the system capable of handling the new concepts of particle-waves was already set in place! We suggest that we should not be surprised to discover that when the limit of quantum mechanics has been reached—the next step will be plainly before us!

From *The Dancing Wu Li Masters* by Gary Zukav, a scientific writer, we gather the following information: "In the autumn of 1927, physicists working with the new physics met in Brussels, Belgium, to ask themselves this question, among others: What is it that quantum mechanics describes?" This meeting resulted in the statement known as the Copenhagen Interpretation of quantum mechanics, which says in effect that it does not matter what quantum mechanics is about! The important thing is that it works. "Quantum mechanics discards the laws governing individual events and states directly the laws governing aggregations. It is very pragmatic." The microscopic nature of "whatever" makes it illogical to presume to speak of individual entities—but we must know about the mass—the aggregate. No one has ever seen an electron or a proton or any of the subdivisions of the proton, but in the aggregate, they behave according to the laws of quantum mechanics (Zukar, pp. 62f).

As matter was broken down to ever smaller bits, the time came when it was not possible to account for them by means of the Newtonian laws of classical physics. It was at this point that quantum mechanics stepped in to take over and has gone on in its mission to identify and describe the particle-waves of energy out of which all matter is formed. At what appears to be almost the end of its road, quantum mechanics has identified six leptons (electron-related) and six quarks (proton-related) which, together with the W and the Zo particles, constitute the fundamental building blocks of matter.

Leptons and quarks are energy forms, of the same substance but functioning differently, somewhat like electrons and protons. They "lived" in an earlier time—much closer to the actual explosion of the Big Bang. This can perhaps be understood by remembering that they represent the sustained expressed will of the Creator-God as he exercised it at the time of the creation explosion. These two energy forms were alike but destined to function in a manner that would result in the formation of matter. It could also be said that they were under the control of the force of gravity, which we have stated was the "line of responsibility" between the Creator-God and the created stuff. Gravity is the measure of the level of attraction between the Creator-God and the Creation, or the "unit of control" exercised by the Creator-God as through his expressed will he creates and sustains his creation. Thus gravity is the ultimate force of Nature, embracing the other three. It is the One that rules them all! Thus the leptons and the quarks are the first "visible" evidence of the Creation—on their way to making electrons and protons, even neutrons, photons and neutrinos, for the purpose of creating the Universe as we know it. It was the peculiar function of quantum mechanics to lead the scientists back to the very beginning, beyond which it could go no farther.

Just as Newtonian physics arrived at a point where it could no longer be effective in describing physical matter, so quantum mechanics has come to a point, the creation explosion, where it appears unable to penetrate deeper. That point, and it cannot be defined exactly, lies in the realm of the particle-waves of leptons and quarks, together with the W and Zo particle-waves. This should not have come as a surprise to physicists, for they had been warned early on that this could happen.

"The Copenhagen Interpretation, in addition to the pragmatic part, has the claim that quantum theory is in a sense complete; that no theory can explain subatomic phenomena in anymore detail. The extraordinary importance of the Copenhagen Interpretation lies in the fact that for the first time, scientists attempting to formulate a consistant physics were forced by their own findings to acknowledge that a complete understanding of reality lies beyond the capabilities of rational thought. It was this that Einstein could not accept. 'The most incomprehensible thing about the world,' he wrote, 'is that it is comprehensible.' In another setting, Einstein stated his position, rejecting quantum mechanics as the 'fundamental physical theory.' . . . 'Quantum mechanics is very impressive,' he wrote in a latter to Max Born, '. . . but I am convinced that God does not play dice'" (Zukar, p. 92).

Thus, while quantum mechanics has come close, there is one step yet to be taken. When it will be taken, we cannot say, but it will not be in the direction some scientists are following today. John Boslough, science editor of *U. S. News and World Report*, has written a biography of Stephen Hawking, thought by many to be the most remarkable scientist of our time. In *Stephen Hawking's Universe* (page 125), we read: "Recently some physicists have come to see a relationship between their work and the ideas behind Eastern mysticism. They believe that the paradoxes, odds, and probabilities as well as the observor-dependence of quantum mechanics have been anticipated in the writings of Hinduism, Buddhism, and Taoism. Quantum mechanics, these new physicists are fond of pointing out, is really only a rediscovery of Shiva or Mahadeva, the Hindu horned god of destruction and cosmic dissolution . . . The god's dance symbolizes the perpetual process of universal creation and destruction. Matter has no substance at all; it is merely the dynamic, rythmic gyration of energy coming and going."

These scientists have come to the point where their minds have been numbed by the concepts which have grown out of quantum mechanics. They refuse to accept the obvious, that the Creator-God must be included in their equations. Like drowning people, they clutch at straws—even Eastern mysticism, which has itself come to the end of the road. They are ready to settle for a philosophical position, a statement to the effect that there can be no answers.

This kind of thinking does not account for the design with purpose and goals which is so evident in the Universe. It is a denial of the obvious! It is true: human resources are not enough. The fact that matter has been broken down into non-material particles-waves or energy forms means that the answer to the problem of the origin of the Universe is not to be found in material things, but in the spiritual, the word of God, the sustained expressed will of the Creator-God. Turning to Eastern mysticism is simply evidence of the futility the scientists feel because they have run out of ideas. Boslough relates Hawking's thoughts on Eastern mysticism: " 'I think it is absolute rubbish,' said Hawking. I looked up from my notebook at him. 'Write it down,' he ordered. 'It is pure rubbish' " (p. 125).

Stephen Hawking is of the opinion that quantum mechanics has gone as far as it can go. "So maybe the end is in sight for theoretical physicists if not for theoretical physics. . .
The point is . . . that we've come such a long way in the last twenty—or fifty—years that one can't hope that it will just go on like that indefinitely. So I think it is altogether possible that we'll either become bogged down and have no more progress or that we'll soon find the unified theory, possibly within twenty years" (Boslough, p. 128).

One reality is the Creator-God. As long as He maintains his expressed will, so long will matter or the Universe as we know it continue to exist. Having come to the point where they have discovered the six leptons and the six quarks and the W and Zo particles, the scientists are standing at the holy of holies, where the energy of the sustained expressed will of the Creator-God is constantly being translated into the particle-waves of which the Universe is made. At this place, if the scientists would simply kneel together with the theologians, hold hands and look "up," they would see what no man has ever seen—the face of the Creator-God. The "unit of gravitational force" would be like a shining light as it draws the particles-waves into "becoming." The Energy of the Creator-God is thus constantly being transferred to the Universe that it might continue to exist. The sustained expressed will of the Creator-God is at work: $E = MC^2$.

Bibliography

The Holy Bible
Children of the Universe
    Author: Hoimar von Ditfurth
    Publisher: Atheneum
    Date: 1974
The Origins of Life
    Author: Hoimar von Ditfurth
    Publisher: Harper and Row
    Date: 1982
Genesis: The Origins of Man and the Universe
    Author: John Gribbin
    Publisher: DeLacorte Press
    Date: 1981
The Death of the Sun
    Author: John Gribbin
    Publisher: DeLacorte Press
    Date: 1980
Beyond the Moon
    Author: Paoli Maffei
    Publisher: MIT Press The library also listed Avon Press
    with a date of 1978
    Date: 1978
The Origin of Species
    Author: Charles Darwin
    Publisher: Modern Libraries Unabridged Edition
    Date: 1936
The Descent of Man
    Author: Charles Darwin
    Publisher: Modern Libraries Unabridged Edition
    Date: 1936
God and the Astronomers
    Author: Robert Jastrow
    Publisher: Norton
    Date: 1978
    Warner 1980

Until the Sun Dies
    Author: Robert Jastrow
    Publisher: Norton
    Date: 1977
    Warner 1980

Astronomy: Fundamentals and Frontiers
    Authors: Robert Jastrow and Malcom Thompson
    Publisher: Wiley
    Date: 1974

The Universe
    Author: Isaac Asimov
    Publisher: Walker
    Date: 1980

The Collapsing Universe
    Author: Isaac Asimov
    Publisher: Walker
    Date: 1977

Exploring the Earth and the Universe
    Author: Isaac Asimov
    Publisher: Crown
    Date: 1982

The Nature of Time
    Author: G. W. Whitrow
    Publisher: Holt, Rinehart and Winston
    Date: 1973

Time In History
    Author: G. W. Whitrow
    Publisher: Oxford University Press
    Date: 1988

Gravity and Levity
    Author: Alan McGlashon
    Publisher: Houghton Mifflin
    Date: 1976

The Nature of Matter
    Author: Otto B. Frisch
    Publisher: Dutton
    Date: 1973

The First Three Minutes
    Author: Steven Weinberg
    Publisher: Basic Books
    Date: 1977

The Edge of Infinity
    Author: Paul Davies
    Publisher: Simon and Schuster
    Date: 1981

The Runaway Universe
    Author: Paul Davies
    Publisher: Harper and Row
    Date: 1978
Journey Through the Universe
    Author: Thomas L. Swihart
    Publisher: Houghton Mifflin
    Date: 1978
The Left Hand of Creation
    Author: John D. Barrow Joseph Silk
    Publisher: Basic Books
    Date: 1983
The End of the World
    Author: Richard Morris
    Publisher: Anchor Press
    Date: 1980
How the Brain Works
    Author: Leslie Hart
    Publisher: Basic Books
    Date: 1975
    Life Science Library
Matter
    Author: Ralph E. Lapp
    Publisher: Time, Inc.
    Date: 1963
Energy
    Author: Mitchell Wilson
    Publisher: Time, Inc.
    Date: 1963
The Cell
    Author: John Pfeiffer
    Date: 1964
Basic Electricity
    Author: Abraham Marcus
    Publisher: Prentis Hall
    Date: 1974
Cosmos
    Author: Carl Sagan
    Publisher: Random House
    Date: 1980
The Story of Quantum Mechanics
    Author: Victor Guillemin
    Publisher: Scribner
    Date: 1968

Coming of Age in the Milky Way
    Author: Timoth Ferris
    Publisher: William Morrow and Company
    Date: 1988
A Brief History of Time
    Author: Stephen Hawking
    Publisher: Bantom Books
    Date: 1988
Stephen Hawking's Universe
    Author: John Boslough
    Publisher: William Morrow
    Date: 1984
The Dancing Wu Li Masters
    Author: Gary Zukav
    Publisher: Morrow
    Date: 1979
The Particle Connection
    Author: Christine Sutton
    Publisher: Simon and Schuster
    Date: 1984
One Hundred Billion Suns
    Author: Rudolph Kippenhan
    Publisher: Basic Books
    Date: 1983
The Universe
    Author: Byron Preiss, Editor
    Publisher: Bantom Books
    Date: 1987
The Solar Planets
    Author: V. A. Firshoff
    Publisher: Crane-Russak
    Date: 1977
The Planets
    Author: Richard Baum
    Publisher: Wiley
    Date: 1973
The View From Planet Earth
    Author: Vincent Cronin
    Publisher: Morrow
    Date: 1981
Theologian Can Accept Evolution
    Author: Philip Hefner
    Publisher: Fortress Press
    Date: 4-20-1983 (The Lutheran)

Evolution and Evolutionism
  Author: Huston Smith
  Publisher: The Christian Century
  Date: 7-7-1982